Tom Cohen

Ecocide and Inscription

Volume I: Black Ops

CCC2 – The Nethercene: Ecocide & Inscription

Series Editors: Tom Cohen and Claire Colebrook

Taking in the vortices and reversals of the 'post' pandemic anomie, The Nethercene sub-series of CCC2 opens the space of provocation for readers during this Potemkin transition by seizing upon and opening problematics of the inscriptive forces and demons before the screen projections of 'consciousness' coalesces – the netherworlds on which these spectacles depend. The Nethercene opens a space to the side of the 'endgame' logic of Anthropocene Talk's panicked turn to claim ever greater 'otherness' in a manner that has sealed the fate of the transitional decade we are leaving. By exploring the laproscopic order of inscription, hermeneutic blow out, referential panic, and Medieval tele-herding that offers distractions and options, this series takes up the remains of reading.

Tom Cohen

Ecocide and Inscription

Volume I: Black Ops

Petrolepathy, Escaped Slaves, Cinemacide
A Tele-Mnemonics for the After-Times

OPEN HUMANITIES PRESS

London 2025

First edition published by Open Humanities Press 2025

Copyright © 2025 Tom Cohen

Freely available at: http://openhumanitiespress.org/books/titles/black-ops

This is an open access book, licensed under Creative Commons By Attribution Share Alike license. Under this license, authors allow anyone to download, reuse, reprint, modify, distribute, and/or copy their work so long as the authors and source are cited and resulting derivative works are licensed under the same or similar license. No permission is required from the authors or the publisher. Statutory fair use and other rights are in no way affected by the above. Read more about the license at creativecommons.org/licenses/by-sa/4.0

Cover Art, figures, and other media included with this book may be under different copyright restrictions.

Print ISBN 978-1-78542-147-1
PDF ISBN 978-1-78542-146-4

OPEN HUMANITIES PRESS

Open Humanities Press is an international, scholar-led open access publishing collective whose mission is to make works of contemporary critical thought freely available worldwide. More at http://openhumanitiespress.org

Contents

Acknowledgements	9
Preambles: In Praise of Malignant Narcissism – or, the Viscous Undertow of Nethercene Reading	11

I. PETROLEPATHY

1	The Viscosity of the Image: Itinerant Photography vs. Cinema in the Era of Climate Chaos	27
2	Roaming Discharges: Avital Ronell's "Catastrophe of the Liquid Oozing"	57
3	Thirteen Ways of Looking (Back) at ISIS – or, "Jihadis 'R Us"	75
4	Cin-Animation: Notes on Hitchcock's "Bird War"	91

II. HAVOC IN THE "PLANTATIONCENE": ESCAPED SLAVES

5	Trackings: Faulkner's Closure of *Ecriture* (*Go Down, Moses*)	113
6	*Sula*, and the Rupture of Mourning	137
7	The Kettle and the Worm: A Note on the Resistance to Climate Change (Faulkner, de Man, Stiegler, Derrida)	153

III. DIOGENES'S LAMP – or, CINEMACIDE

8	Arche-Cinema and the Politics of Managed Extinction	187
9	Did an "Anthropocene" Even Take Place…?: Notes on an Image in Spielberg's *A.I. Artificial Intelligence*	215
10	Exploding Omnibus – or, "This Ain't No Fucking Movie!" (on *Bus 174*)	237
11	Notes on the "High Trumpocene" – or, the Discreet Charm of Climate Autocracy in the Time of Cascade Events	259

Works Cited	273
Notes	279
Images	295

For Sigi Jöttkandt

Acknowledgements

The essays in this volume have been informed by many interlocutors and provocateurs, to whom I'm deeply indebted. Among the so-called living, these include Sigi Jöttkandt, Claire Colebrook, Jonty Tiplady, Helen Elam, Daniel Myerson and Eduardo Cadava. Sascha Gebauer, Henry Sussman, Timothy Clark, Avital Ronell, Gabriela Nouzeilles, Denilson Lopez, Jonathan Hart, Min Zhou, Wang Fengzen, Wang Guanglin, Martin McQuillan, Erin Obodiac, Liana Theodoratou and Alexander Strecker have all been critical to this project at different junctures. I also thank professors Myung Ho Lee, Kim Jonggab, and Kim Mijeong for productive interrogations, and Barbara Herrnstein Smith for her many and long-standing provocations. My special thanks to David Ottina, without whom there would be no Open Humanities Press.

Versions or segments of these essays have appeared in *Discourse*, *Oxford Literary Review*, *Parallax*, and *Boundary 2*.

Preambles: In Praise of Malignant Narcissism – or, the Viscous Undertow of Nethercene Reading

> We have a task before us which must be speedily performed. We know that it will be ruinous to make delay. The most important crisis of our life calls, trumpet-tongued, for immediate energy and action. We glow, we are consumed with eagerness to commence the work, with the anticipation of whose glorious result our whole souls are on fire. It must, it shall be undertaken to-day, and yet we put it off until to-morrow, and why? There is no answer, except that we feel perverse, using the word with no comprehension of the principle. To-morrow arrives, and with it a more impatient anxiety to do our duty, but with this very increase of anxiety arrives, also, a nameless, a positively fearful, because unfathomable, craving for delay. This craving gathers strength as the moments fly. The last hour for action is at hand. We tremble with the violence of the conflict within us, – of the definite with the indefinite – of the substance with the shadow. But, if the contest has proceeded thus far, it is the shadow which prevails, – we struggle in vain. The clock strikes, and is the knell of our welfare. At the same time, it is the chanticleer – note to the ghost that has so long overawed us. It flies – it disappears – we are free. The old energy returns. We will labor now. Alas, it is too late!
>
> —Edgar Allan Poe, *The Imp of the Perverse*

Perhaps we can summon Poe for a scan, and call to innocence, as to the big picture on climate chaos and extinction imaginaries today – the "great dithering" (Haraway) of the last two decades, including the moth-to-flame detour of "Anthropocene Discourse," the discreet barreling through tipping points, the fantasia of techno-salvation…. One might wed this "craving for delay" to recent modeling probes that determined *any* planetary civilization's tech explosion commandeering energy transference would have a maximum of a thousand years before burning out and disappearing. Seems like a lot. One closes one's eyes to savor: if it were not *these* infantile billionaires, it would be others; if not these centuries-old political improvisations, some other (the so-called Fermi Paradox rendered unparadoxical). That suggests that this time window or decade is exceptional: when irreversibility and cascade events trigger, altering not the chance of outmaneuvering mass extinctions but biodiversity for millions of years. Let's just suppose this cannot be fobbed off by blaming some mutant flavor of the big C (Rogue Capitalisms, neo-medieval, tech-feudal post-Capitalisms), nor one or another national grouping to point to and expect response or culpability (Big Oil will "Big Oil"). And let's assume for a moment that buried within "our" inability to respond *in time* to what is mislabeled the *climate crisis* is something akin to, say, Poe's *Imp of the Perverse*. We *might* seem to have chosen otherwise, but we have actively and passively chosen to "gamble on climate change," to overshoot and *see* – or, rather, a techno-oligarchist breakaway caste does, confident they can manipulate the timing like a tourniquet for the right, purgative, timeline results, whose imaginary crystallizes not in Arctic condos and underground Alpine cities, but "Mars." Especially after having watched the movie in repeated variants, one may now finally *see* for oneself, enter the video console's embattled wastelands? If "we" are going through the acceleration tunnels, is there an imp of the perverse driving a will to experience this? The collection on ecocide and inscription might be a remote rogue cousin to Yuval Noah Harari's recent *Nexus: A Brief History of Information Networks from the Stone Age to AI* (2024), wherein the attempt to give a new "story" to the human or hominid arc – and its techno-future – turns to map the whole as a history of mnemotechnic networks flat-lining into today's dark and looming merger of "infotech" and "biotech." Along such an arc, these sketches perhaps fill in what Timothy Clark chose to identify as an "inhumanist" school (of two), by which is meant recalling that the techno-digital orders of inscription, like that of writing and the book, are themselves not human in any familiar sense of the term – requiring implements, surfaces, marking systems, mnemonic institutions, mass dissemination technologies (cave walls, feuilletons, screens).[1]

As its portmanteau title, *Black Ops*, insinuates, there is a collusion between an interrogation of hyper-blackness and the interface between optics and invisible targeting operations. It follows, in a stupid way (not meant negatively), that if cinema were the stalking horse, implant, and proto-networking of AI in its pre-digital mode, it also stood as the evolving locus of archival management, culling, "reading" to come.

Hiatus I. *"Anthropos" – We Hardly Knew Ye...*

> More black than the blackness constructed to justify slavery in the era of colonialism...More black than the black market today, where human beings, together with drugs and arms, continue to be traded as illegal commodities, whose general investment in the production of value is enormous but whose slave, unpaid, or low-paid "dirty work"...is not visible because it is not socially represented...oil...is not a master, but a kind of ultimately inhuman black slave.
> —Oxana Timofeeva[2]

> These are notes on blaccelerationism. This portmanteau – binding blackness and accelerationism to one another – proposes that accelerationism always already exists in the territory of blackness, whether it knows it or not – and, conversely, that blackness is always already accelerationist. It is my modest proposition that activating this blaccelerationism serves to articular a necessary alternative to right and left accelerationism.
> —Aria Dean[3]

"The blobjective earth is nurtured by petropolitics," observes Reza Negarestani. Perhaps, but it is sustained by petrolepathies that orchestrate aesthetics and visibility by effacing its ubiquity and agency.

Anthropos likes to hedge his bets. No one wants to give "anthropomorphism" up altogether – unless there were a more abyssal narcissism still that precedes that, before "face" is posited or seemingly contracted, self-archived and tagged as a debt, a promise, an assurance, a con. And even then, no one wants to sacrifice certain things: their energy expenditure, their smartphones, their investments, their credit with diverse mafias, or the perks of "personhood" held hostage to a system of privileged exchange – exclusion from which, in a managed mass extinction event, triggers new zones of disposability. A consolidation, a breakaway (financially engineered, techno-eugenic, species splits), the start of the cascade events, encompassing migration into screens, a narrative of return to oil as necessary if not open-ended... ?

These writings do not project an escape from the "Anthropocene" trap, and turn against, as contretemps, the prevailing panic turn of legacy-theoretical idioms seeking to reconfigure, and preserve their investments, negotiating the implications of tipping points and progressive uninhabitability openly on view. Rather than assuming one is trapped by anthropomorphism, for which the escape is to break out, move more to the outside and identify with diverse "others," supposing access and so on (racial and cultural others, then animemes, microorganisms, the inanimate, en route to Meillassoux's ultimate regresso-trope: "the great Outdoors"), reading here asks first to melt back into the order of inscriptions themselves, instantly apprehended if not appropriated by social media and "political" algorithms, mnemonic hacks, coding. These pieces instead track a thread that leaps between writing, cinematics and climate chaos – and the ghost of "AI" (whose data-center needs are calculated to gulp 19% of US energy consumption in a few years). "Now," these two – AI and climate extinctions – appear tied together in a race, in a ticking-clock *mise en scène*. A double vortex.[4]

These sketches at times track the pre-mimetic mark, the line, sound, the sheer anteriority of script and acetate and algos, as miming into the paradoxes of oil, or carbonic ink, the pharmakon becomes digital poison, the leakage across cells like the microplastics now lacing inhumans and fish alike. Subtending light and visibility, it also testifies to the explosive, hyperbolic energy expenditures initiated with the discovery of oil in Pennsylvania by Edwin Drake (see "Drake Oil Well" museum).[5] By chance, this Drake is a shadowy forebear on my mother's side – whose family goes back to pre-revolutionary days and claimed, along the way, a dubious blood roster of American performance artists (Daniel Boone, Henry Clay, William Tecumseh Sherman, and Drake – who, displaying a comic family trait, unloaded the thing to a guy named Rockefeller). Perhaps some lingering debt here induced me to track this toxic black stuff, this "stored sunlight," in the time of climate chaos – this cannibalistic prey of (human) life feeding off all past organic matter, all previous life forms, burning off into an atmosphere choked and turned into a suicidal trap (twenty-first century "Earth"). Toxic to touch, it – if it were an "it," what we name "oil," now suffusing plastics, foodstuffs, bodies – guarantees visibility yet is itself removed from the visible. It is contraband to discourses at play on the surface, where, occasionally, a damaged ship or explosive "leak" will draw cameras and attention to it until that can be sequestered. It passes under, in rivulets of the Nethercene plains, like blood, and turns up as ink, poison, pre-mimetic figures of blackness, a zone of sheer anteriority unaware that it returns from some past perpetually, in a backloop.

There is a curious effect in reading as the bricolage of aura and personification itself recedes, as "anthropomorphism" is at once put in reverse and becomes inflamed and faux manic, regressive, stupefied. Herded memory programmes are swapped, SIMs installed, hacked and whited out with sling terms mesmerized by the time of climate chaos's broad panic of reference. The occluded or deferred zone of climate disarticulation, to which all current species are subordinate, is indissociable, burrowed in gathering geopolitical storms and the migration into screens. With mass extinctions and uninhabitability spawning, nothing matters more than Russian grudges, Chinese megalomania or American stupefaction and anomie converging to grab lunar fields, fry grids and satellites, render all prospect of biospheric containment out of reach, "too late."[6]

All of this today informs and frames and inflects any reading expedition, alert or not – whose eventual critical readers may be that hybrid AGI cohort to come that will no doubt manage and cull all archival detritus. They may be fascinated by what informed our dawdling. Being without "bodies" and able to digest data streams and read (in their way) instantaneously, they may note how much received hermeneutic programs were complicit in the delays and deferral of the opening decades of this "century," guaranteeing the accelerations "we" witness beginning. While buffeted by climate threats, we experience these as if rewatching a film, willing to go over the edge to see, finally, what we've been digesting and entertained by on screens.

Now that we hover in the *after-times*, one can track other formations at work the Nethercene. If there is a malignant narcissism to the double-pharmakon of fossil energetics and its incremental demands, it adheres not to the sociopathic composition of "networks" in the current moment of climate rupture – that of the slow rollover into the cascade sequences pending and deferred in the imagination even as it sweeps through diverse ecological regimes.

"Anthropomorphism" as a term is cast about as something we can outwit or even suspend, as when we identify it in another's arguments. It seems a tautology seldom defined so much as assumed. What it assumes is a projection of human qualities onto things and entities that subject them to our extractivism and economies: they assume that this given, "we anthropoi," can be exported into the inorganic and merely animate, since it is "we" doing so. It is, or seems to be, a trope. As such one might imagine that the mechanics of tropes would be one of many zones explored by sophisticated means – yet, on the contrary, this is avoided, as if some unpleasant secret lay within its premise.

The opaque premise of a universal anthropo-narcissism covers a perpetual non-site where the effects of voice and "consciousness" appear to emerge, now simmed by AI Chatboxes and pre-captured by algo streams and mnemonic implants. Yet, this "consciousness" that is more universal-seeming than the hominid form itself is contradicted by anthropomorphism not being a figure or a "trope" itself. It retains the perspective of the vast mutations of hyper-materialities before and outside trope or artefacts. Anthropomorphism, it turns out, precisely bars any universalizing notion of man, since it can be constituted, only, as an exclusionary totalization.[7]

What I track under the premise of *petrolepathy* or more generally *back ops* (by which "optics" is heard) returns to a formation I focused on in Hitchcock's early, or British, work. There a black sun, black cat, black dog, and so on traverse the screens with strange agency; I found myself drawn to where the figure of blackness itself appears to precede the binary it is wedded to, accumulating a switchboard of threads and networks encompassing race and escaped slave figures, liquefying in modes beneath one or another tropological front to mark an irreducibly carbonic trace, the specter of mnemo-"materiality" before reference is imposed, a non-site that (continuing this riff) leads through the paradox of how "oil," as if to say itself, evades entering the representational domain, or screen, which it subtends and pervades in all machinal and video-matized artifacts and backloops. It – making photography and cinema possible – is representational contraband, a waste product (Hitchcock allies his black sun avatars to turds or excrement repeatedly), toxic, and not to be looked at – apotropaic, at the core of a polar vortex. The leakage that I trace marks itself while tied to the vortex of climate and biospheric mutations.

What interests is where the order of inscriptions dissolves with this liquefaction. It suggests how the totalization of hermeneutic spells led us to glide over tipping points we knew to be irreversibly set to undo the biospheric template our life-forms happened within. Lingering in the question of what arrives as inscription is the question of disinscription, of altering the tropological programming and algorithmic neurohacking to conjure a reset, what amounts, like genetic engineering of the individual body's production of a new "present," to noetic intervention. All with the time of mass extinction events as background.

These sketches thus emerged in the seances of reading in the era of climate chaos, now passing to that of managed species splits and delayed extinctions. Unexploded ordinances on the seabed of the Nethercene. They inhabit and are curious about the impending retirement of Anthropos 1.0, that is, of what is now better labeled we "Neanderthropoi." This leads back in loops to a hyper-blackness dissociated from any binary, allied with technics itself. Such looms within all letteration and digital streams as mnemo-technic transmissions, delayed, auto-recorded, pre-edited, disbursed. Here a black hole of retro-architected transmissions spins. Such arrives, for us, in mnemonic circuits and technologies, accelerant from cave painting and ur-cinema across writing systems into industrial-era cinema and the digital overwrite of screen "life." And AGI as the remedy or *pharmakon* "of" organic life. All anarchival systems would be precursor modes of AI. Thus, the difference of LLM, large "language" models, built out by ingesting, modeling, and regurgitating "all" writing – at risk, when feeding troughs are shut off, of cannibalizing itself. The output, a SIM "voice" or persona, alerts us to a composition by external citation – of intellection, of personhood, of (ambiguous) sentience itself. It can then promote tutelary identification, friendship, psycho-erotic bonding to the point of mating. The AI Chatbot weaves citations to posit and deduce, from which its "I" is retrofitted as a subject mirage. Like a certain hominid.

When Hitchcock shatters this enigma of a black sun into myriad, attacking, black points in a sunless sky (animeme birds); when Faulkner drifts into an identification with the runaway black slave of the ante- (and post-) bellum Plantationcene, and does so on behalf of script itself; when Avital Ronell draws our readerly eye to the back ooze flowing from the dead Mme. Bovary's mouth; when Sula indexes a blackness preceding face or race – we seem, I would posit, in an inter-networked ensemble of leakages, petrolepathies, black ops, and hyper-material regress to the site where anteriority and inscription and biosemiosis converge and bifurcate. This is of course also in play when something hidden from the eye yet responsible for generating visibility and light and biospheric overshoot, what we call "oil," erupts into view in photographematic screens or displays.

The "immense regression" (Stiegler) we witness in psychic, "political," financist, meme-religious and intra-psychic terms testifies to a "panic of reference" underlying the era of climate chaos accelerated beyond scale, range or epistomography and, in turn, migrated into screens and algo storms. None of the terms, cyclical expectations, living premises (climate chaos, resources) or hanging catastrophics (deferred over decades but in place) dissolves the legacy premises of linguistic use and conceptual maps. Beneath it, an unmooring of referential anchors, inscriptive radiation, the white holes of info-saturation, the reign of deferral until one is irreversibly accelerant, as matters and wars re-align beginning with the premise (or fact) of it being "too late."

Hiatus II. *In Praise of Effluvia...*

Let's say there is a pause in transitions, between epochs and referential settings, when the giant wave is curling beautifully or the maelstrom has just shifted from a slow, entropic roil to begin its funnel accelerations. Such a hiatus might be imagined in the undesignable transition to a post-tipping points, twenty-first-century mise en scène of a progressively disinhabitable biosphere. Such a pause or tremor would seem, today, to both epitomize the banal underbelly of the Anthropocene imaginary and to revoke it as a passing meme we living have used as a Rorschach of sorts, a scratching post, or found consolatory. Is there an underexamined intimacy between writing itself, a carbon trace in its ink imprint, and extinction logics that requires of the speculator a sort of de-reading? Are there moments alert to this all along, or sites where the skein of figurative language we tool about within melts back into its molecular formations, leading the eye to track what constitutes one of its necessary blinds or dis-enframings. Thus, a sort of black ooze before alphabetic characters or "light" – what Oxana Timofeeva has termed "ultra-blackness," absorbing not only the racial binaries and decomposition of the late Anthropocene but a signature of carbon? The question bubbles up when the slave order of technics manifests itself or begins to leak – like, Avital Ronell points out in *Crack Wars*, the black bile leaking from the dead Bovary's mouth. Ronell calls this, wandering off all representational grids, the "catastrophe of the liquid oozing'."

The shift referenced alters temporal chains. It creates all manner of defense formation and power grabs amidst the decoupling of material from cognitive infrastructures. In place of the projection of an open future for the "arrow of time" to seek, a different calculation: how to stretch out this closing time window, now on automatic acceleration, irreversibly, so that some putative architects or techno-elite can plan a migratory escape plan: techno-evolve (in time), and get out of Dodge (Mars, anyone?). This would displace an assumed open futurity with the model of stretching out a known terminus, limit, exhaustion or closure – uninhabitability, say – not to beat extinction logics but to slow it down enough for a certain caste to emerge. When today's diet of superhero films shift from being post-apocalyptic to openly siding against us, affirming extinction as just, and identifying with whatever would replace us, we are being trolled.

The lack of "aura" for a black ooze which generates "light" melts back from alphabetized screens into a pre-mimetic zone. It evokes oil along the way, not encoding blackness as the abject or out of sight and off-screen but as a core technic from the borderlands of the organic / inorganic.

I want to pair a certain black light-writing with the petrolepathy that lies below. I pivot from oil to the escaped slave from the white plantation order of hermeneutics, from face, from a "whiteness" tied to extinction drives which the writing identifies with. The tie between each is a petrotelephathic zone incapable of translation back into the mimetic protocols which sustain the illusion of archival continuity and plantation hermeneutics, and initiate an irreversible dismantling of the still automated hermeneutic patching up that tends to make critical protocols complicit in the enframing algorithms of extinction economics broadly: what if, rather than architecting new "others" (racial others, then animals, insects, microbes, stones and anointing oneself a new ethics to come), one flipped the switch. That is, not reflexively reaching outward but back into the pre-mimetic storm systems from which these referential programs were installed? This is a zone both conjured by Derrida and, as if with another hand, passed over to generate a communally binding weave of affirmative "deconstructions" designed, rhetorically, to initially deflect the abjection of de Man and Heidegger from coming for himself. Climate chaos and ecocide is a zone Derrida left unpursued. What if the dominant reflex in Anthropocene Talk were still wedded to archaic mimetic programming which cannot but accelerate the vortices, rather than break with a dawning anaesthesia paralleled by digital totalitarianisms hard and soft. I would pose all of these contests in relation to what I have called a "war of inscriptions" – a war that coding, algos, bot swarms and infant AI have moved to troll and preempt. This may be best heard by those who "live" on the metaphysical shorelines, exposed in any of a number of senses, yet who have a thirst or responsibility for what roams outside the inundated mimetic window.

Thus, the now iconic *Trump*, who at once would restore the reign of white Anthropos and accelerate the carbonic tipping of the current regime of life forms. For this gold-coiffed caricature of the old western Anthropos, de-extincted – voracious, hyper-extractive – we witness a creeping subtext. It coils behind a breakaway caste of .0001 percenters, controlling AI transformations and digital totalitarianisms. That is, a digitally engineered, massive transference of wealth and resources predicated on a self-executing species split with an eye to the future calculations. One might update Nietzsche for a newly vulgar, twenty-first-century pseudo-epoch of "post-truth" distraction apparatuses: not man would rather will nothingness, than not will – but rather, Anthropos is compelled to the ultimate trick of mastery, irresistibly it seems: to at once engineer and survive his own extinction, or that of the old, mimetic, credulous, genetically messy and un-enhanced Anthropos 1.0 version of the past 50,000 years or so. It appears more curious than ever to conjure this liquid blackness outside of mimetic hermeneutics, before and outside of alphabeticism and photographic prints, closer to the Minotaur ur-site in its screen labyrinths from which inscriptions appear set, coalesced, displaced.

Which returns one to the value of a certain malignant narcissism. One might conjecture that in the order of psycho-poetics, a culture of extinction addicted to mimetic premises might regard as malignant what breaks this contract, perhaps on principle – so that, instead of the rotisserie of standard images, tropes and affects, of referential securities, inscriptive orders melted back into the pre-phenomenal pipes and reserves.[8]

Hiatus III. *Naufrages of the Nethercene*

I mused, in reflecting on Sigi Jöttkandt's *The Nabokov Effect*, by conjuring a Nethercene zone of mutating inscriptions and cognitive channeling that decouples from the Anthropocene miseries and noise. This, as cascade events arrive and mass stupefaction self-feeds in digital streams. Such a zone decouples from numbing reactive vortices and tired scripts not because the "Anthropocene" moniker was officially revoked by the nervous geologists who launched it. It decouples to track or witness a transition that curtails and alters the prospects of life forms on this planet, together with their definition (non-organic "life"), alters cosmic neighborhoods. I invoked a Nethercene likened to the deep sea, with its mandibled denizens and repository of fallen surface debris. I likened Jöttkandt's reading expedition to one "at home among octopi." But the seafloor is but one of its alternative milieux. One can visit the Nethercene lockup, where cells are improvised to contain the inert, and one can visit the cyber deserts. But before leaving the Nethercene deep or its diverse "karaoke" rooms of sorts, one is aware of other, less salutary debris or artifacts. Oil rigs, leaks, dead zones, acidic or oxygen-free. So incessant can this gurgling be when proximate that one is compelled to mark, suppress, be overwhelmed, or otherwise account for it in some capping manner. Yet the gurgling proceeds.

Catherine Malabou has deployed the figure of plasticity to explore a neural liquefaction of trace, one which, though indexed to the mimetic scientisms of brain-mapping, looks back on something like "deconstruction" as closed – speculating on an era in which writing as cipher is deposed or withdrawn, placing Derrida in a Roman-legion, tortoise formation with an array of recent male philosophic agents, those unwittingly defending against the liquefaction and complicity of "philosophy" caught in a self-maintenance mode. Other liquefactions – a melting of cell walls, graphic iterations, glaciers – can be diverted to logics which the "brain" cannot capture. If an era of "the book" subsides, with what Malabou suggests is the end of the reign of "difference" (as LLMs and AGI, gulping all archives, usurp tele-mnemonics), it is not only due to mass digital transcription and totalizations.[9] It echoes the way that ink would be a carbon implement which – the premise of "life" forms – passes into the telepoloi of oil and artefacted "light," of hyper-energetic transmission.

Such is affiliated with what Oxana Timofeeva has named "ultra-black" in excavating the lurking agencies, still, of oil. Here a blackness beyond black and, technically, before "light," harasses the construction of visibility itself. If the trend in critical efforts has been to stage a negotiated opening to an endless series of "others" (the abjected cultural "other," but then animals, stones, "objects," whatever can be traded out while maintaining the control booth), Timofeeva goes into reverse – melting back toward that which precedes, energizes, inscribes, and would reclaim the artificed light and screens of the noosphere itself. If the totalization that we witness today in both digital encirclements and mnemonic herding is generated from inscriptions, installed programs and the trolling farms that ape them, the question of this netherworld to the "light pollution" of our attention conjures a site or non-site where inscriptions, as such, would be modified, as in genetic engineering, liquefied, replaced (or digitally diffused). These pass through an underbrush of any "Anthropocene," a constitutive blind preceding and now defining visibilities.

It remains one of the lacunae of Anthropos's narrative rise from the Greeks, his installation, that the so-called cynic Diogenes of Sinope waved about an oil lamp in the daylight, unable to locate an Anthropos, an honest one, such as Aristotle had launched and defined by exclusions. This is taken banally for his claim not to find an "honest" man, but the emphasis should be on the Anthropos himself, engineered by Diogenes's contemporaries: he cannot locate a true or honest Anthropos. That curious lamp also hints at Anthropos's carbon-tricked destiny at the origin of cinematic projections, the insertion of a technic lamp over solar light. Ignoring the misprisions of the term "cynic," here, what is useful to mark is that he is nonplussed by the fabrication of the Greek Anthropos at its launch, saw that as fictional, a controlling definitional. He mutely puts the "history" that Anthropos will spawn in question as a narrative. Raphael places Diogenes as outside in the *School of Athens*, the only horizontal figure, lying down on the steps, reading – the only reader in the scene.

The Anthropocene appears conjured by a cannibalistic feeding off the decayed waste of all preceding life forms (fossil fuels). It's exhausting aeons-worth in decades, emptying the carbon netherworld back into the atmosphere and sunlight again, tipping the climactic balance of the biosphere against its contracts of residence (a Holocene that may have lasted another fifty thousand years on its own if not abrupted). Cinema and the photographic image, premised on the oil era, rarely reference this dependency overtly, oil itself being formless, without face, without light, "ultra-black," the very opposite of charismatic. It is deemed toxic to touch, is viscous, permeates perceptual programs and preserves invisibility. It enters the frame or visibility only when draped over or in contact with dying animal life forms at spill sites.

Perhaps an errant set of interventions is already overrun, given the kernels of these pieces span a decade or more – but I seem to have started and ended with the same questions throughout the before-time (the bubble detour of Anthropocene discourse), which is simple enough: given we have overshot civilizational and energetic thresholds passing tipping and opening upon cascading networks and roiling micro and macro extinctions, we can afford to decouple and consider what being too late and passing tipping points initiates for us (let's leave that pronoun pending). It might switch PoV's and ask how these earth systems look back at "us." Whether a different "materiality" (conjures) itself outside of legacy binaries (new materialisms, hyper-objects), which must keep the eeriness of carbon in mind as underlying inscription technologies and what had been "writing" (ink). So these sketches are from the before-time, before the cascade effects have fully manifest, at least their beginning, or not.

For myself, this path leads through select concerns of "deconstruction," which nonetheless seemed curdled and inept after Derrida's death due to Derrida's choosing to ignore the one catastrophic underway, not marked in the twentieth century and to the side of any "affirmative deconstruction" he felt compelled to construct to resist a threat of cancellation (following de Man and Heidegger).

Writing as if in 2025, there seems a light-headedness or vertigo in the air. There are innumerable fresh and compounding accelerations and responses in play, as new delays and denialisms and regressions (back to oil, etc.) extend themselves. It is entirely banal to note – so unimaginative – the power of techno-oligarchs and neo-fascists converge with economic techno-feudalism and war as wobbling systems grab for future resources. The view from the top of this rogue's gallery of hyper-powered deciders fuses with a heady cocktail of gangland nihilism, techno-eugenicist fantasies and a working assumption that the genetically messy masses inherited from the Holocene, oil and hyper-population will be slowly retired by climate chaos, as a new hybrid human form takes over and migrates off a ruined earth to a Mars joyously accoutered with potato farms, a Starbucks and a Trump casino. So, let us muse: if one were to track changing sentiments and eco-political black holes, we would have to say we are already skipped forward a couple of tracks from the last scans. The Covid pause seemed to have marked this, albeit inverse, entry into a field of anomalies and incipient vortices expanding on auto-pilot. *Climate comedy.*

Oil here operates as a kind of "contraband" within the circuits of signification, a black market of meaning trafficking between the graphic and the geomorphic, the symbolic and the symptomatic, the ecologic and the ecocidal. As the secret key to our fossil-fueled modernity's ecocidal logic, oil has been the dark matter lubricating the gears of our global economy, geopolitical imaginaries and apocalyptic fantasies. From colonial quests for "black gold" to the racialized distribution of toxic exposure, oil and race have long been entangled in a double helix of domination and disavowal, yet oil also exceeds and recedes from social identifications, from racial or gender binaries, from and into its own black pool where the debris of writing networks and archival effluvium melt back, in awe at the captivating and ruinous spell that, cast, now runs out and overshoots on auto-pilot.

It is a feature, not a bug, of 2025 that a return to oil, after and in the midst of various broken promises by nations left and right, is heralded and become accepted as overriding the climate urgencies routinely shouted out by the U.N. Secretary General and his feed lines. And it makes a peculiar public acceleration and occlusion with the installation of the Trumpocene as now normative, of Trump-Musk clarifying the dismantling accelerations and the species split underway dovetailing with the flickering twilight of duration reading (and perhaps alphabeticism).

■ ■ ■

Note: If we disperse the long focus on "the" Anthropocene as a detour and distraction, a sort of cat scratching pole for all legacy investments in political and humanistic "universals" to update their software (it remained anthropos-centric after all, despite all edited modifications and counter-names), we are not surprised by what is now disclosing itself, and which is tracked in these essays as the lag effects of a hermeneutic pandemic that degenerates in the era of discredit into conspiracy weaves and trolling propaganda that can't be bothered to mask itself much – nor needs to.

What emerges in place of any "Good Anthropocene" imaginary is not simply its dark inversion – the High Trumpocene – but rather the installation of an unprecedented onto-political apparatus that transforms climate irreversibility into the very precondition for engineering posthuman succession. Trump manifests as the first true "climate president" precisely by operationalizing this irreversibility: the systematic excision of climate discourse from institutional lexicons doesn't merely repress but actively weaponizes environmental collapse as the ground for species bifurcation. This regime grasps – indeed, predicates its entire operating system upon – the impossibility of genuine climate mitigation for any but a financially-engineered survivor caste. What crystallizes is a triadic assemblage that metabolizes catastrophe into operational advantage: evangelical rapture-seekers converge with petrostate kleptocrats and Silicon Valley posthumanists in a power formation that accelerates breakdown to facilitate the "retirement" of current iterations of Anthropos. *Trumpocene* thus names not mere reaction but the emergence of a terminal political rationality: one that exploits the foreclosure of mitigation to split humanity into a biotechnically-enhanced elite and an obsolescent remainder, consigned to neural-digital screen migration as their biospheric substrate deteriorates. This is not climate denialism but rather the tactical deployment of ecological point-of-no-return as the material condition for executing a species bifurcation that finally "retires" the messy Holocene phase of human evolution. The High Trumpocene thus instantiates nothing less than an extinction-management protocol masquerading as politics – one that transforms irreversibility from obstacle to instrument in engineering Anthropocene 2.0 succession.

I. PETROLEPATHY

Oil

Hypothesis: what we call (and abject as) oil not only suffuses terrestrial organisms (microplastics) but inhabits representational and figural networks – particularly in the tele-thanatographies of photographics and cinema. Such a petrolepathic switchboard remarks itself within and outside of any frame, binding "invisible" (too visible) semantic orders and, now, the rise of A.I. and algorithmic capture to this... leakage.

1 The Viscosity of the Image: Itinerant Photography vs. Cinema in the Era of Climate Chaos

Why does oil itself – about which the energetics of the current acceleration and the engine of biospheric mutation and climate chaos, what give and take "life," are themselves decayed biomass called "stored sunlight" – never find itself represented in writing, cinema, the photographic image? There are exceptions, peeks and surfacings, but otherwise, you say, there is nothing to see: a black pool or blob, unless attached to and altering what it drapes itself about. Toxic to the skin and hence to touch. One can trace its presence in the idea of "itinerant photography" and a certain war of usurpation it briefly tried to stage against cinema, against Hollywood, in the example of Robert Capa's encounter with Hitchcock – and this around the set for *Notorious*, a film set too in Rio de Janeiro.[10]

I

Does the era of cinema overlap with the era of oil and hydrocarbons in a way that frames and fashions visibility? Is there a link between the representation of oil, if it breaks into visibility itself, and the era of climate chaos? This gives itinerant photography, which Eduardo Cadava has invited us to address, a key, if mutating, role – claiming an ultimate embeddedness in "life" (usually war), a hyper-mimeticism. Its movement has a strange relation to cinema – one which maintains a certain torsion or war between technics. I will narrate it, though, as a face-off of sorts between Robert Capa and Alfred Hitchcock. One touching both on oil and on Rio. If cinema as a term implies transport, and machines of transport require oil (and the projector's artifice of "light"), is there a relation between the photographic image and what is today called the "Anthropocene" era, the era of mass extinctions? I will pursue these associations – that of a secret war between photography and cinema, and the invisible place of oil in the frame, which underwrites the epoch of machinal cinema, as the excuse to move through two dossiers. I will put each in parallel dialog with the other. One is that of the Gulf oil "spill" and its representations in web-platformed, circulated photographic displays, as a drama of enforced invisibility, and the second, what photography and cinema think of oil as their unrepresentable premise, from which a certain predatory visibility ensues.

Yet where would oil, which may in fact be the hyposubject par excellence, remark itself? We must stop to even bring it into focus as an "idea," what, were we in a Grecian world, would have to have a god named on its behalf – "stored sunlight," black gold, contaminant to touch, excrescent, the remnant of organic biomass, or the living, seeped into a netherworld that, when released and burned into channeled energy, overtaxes the upper spheres and returns the cycle to itself. Outside, like waste, it is toxic, yet sought like cocaine. It is driven or pulled up from an underworld of "past" organic life (Melville's "Moby Dick" – only unanimated, ink-like, carbonic and not white). In this, any modern present rapidly feeds off aeons of past organic "life," that is, of stored sunlight. Sheer anteriority, plastic or amorphal, it is neither subject nor object, neither form nor fluid, neither organic nor inorganic quite. Black, it absorbs light or gives it, redistributes light as surface, takes the form it drapes itself about (reclaiming, smothering, still animated ones). As it is not strictly an "it," we can say that there is no "oil" as such strictly. It in itself hints at what Eduardo Cadava sees as the black or prosthetic light-writing of photographesis. Seeming to be purely exterior, it courses through the whole, on- and off-screen. It is pure technics. After the Gulf oil "spill," BP spread dispersant chemicals widely so the globs hung below the surface (even if toxicity was enhanced), to avoid the eye or cameras; BP bought up local Gulf police forces to ban cameras, and universities to silence research on the silent spread within life chains and economies. This cannot be dissociated from the visible and what would be made invisible, obscene, criminal, not to be mentioned. It penetrates living chains, encroaches, enters cells and bloodstreams. One can speak of a gulf within oil perhaps that makes its reconciliation as a dead zone spawning semiotic "life," indeed, the noosphere as we know it, seem unreadable – and all "light," now artificed and hence black light in fact, eclipses the sun, exposing it too as a pyrotechnic complex, light itself a matter of alternating waves and frequencies. At its extreme depth, oil spawns speculation of abiotic oil – not of mere organic matter, but of a stain at the source of production itself, escaped from the cycle of life-death, making all engines run with no one time. Another culture than ours might have paeans to oil, temples perhaps, no doubt celebratory media, rather than being maintained out of sight and touch, without face or light, except perhaps in the plastics, celluloids and, now, GMO products we digest.

Sharing the same parentage, who would not expect a war between two modalities and technologies that, intimately interdependent, appear diametrical opposed in their temporal claims or vectors? Has there all along been a certain war between photography and cinema,

whose temporal logics appear to cancel one another's claims to either simulate and totally arrest animation (say, as metonymy and metaphor appear to do within rhetoric)? What are the stakes of such a war in the era of climate chaos, which is the era of oil, the "great acceleration," techno-genocides and extinction events? Has the Anthropos of the

Anthropocene era (if one subscribes to this nomenclature) been a cinematic product or event, and if so, or not, how can we get a shot of or at him – in perpetual artifice and perpetual movement? We are better off approaching these apparatuses not as tools of our invention but as early para-organs, generalized precursors of the smartphone mnemonic-limb of today.

II

For us, we ourselves, oil arrives more or less anthropomorphized, mediated and effaced at once. In Chevron's famous logo, we pass directly to "human energy," a claim that leaps directly into our anthropoid reserves or digestion tracts. It – or the world-trashing multinational that fronts it (think Ecuadorian Amazon) – is good. And in the mesmerizing example of the McDonald's hamburger left to decay in the air for nine months, in which nothing occurs, no bacteria or disintegration at all (indeed, no microorganism will touch it), we see that oil's derivatives (and plastics) now constitute food in fact, ostensibly "human energy." And there is an erotics of oil I will not touch on here; it is much too interesting, all these lubricants, but we'll bookmark

this with the latest James Bond girl redaction – in which Goldfinger's gold-dipped nude of the 1960s returns as a blackened, oil-dipped beauty, politically reinscribed for the day, the anti-meme to all those blackened, flightless birds and turtles the camera picks up after "spills." (Today the naked female corpse might be covered and perforated with… data?) Which may be why, even here, as we assemble reading together stills in a sequence featuring persons, faces, one after another, there is something (I mean this gently) nostalgic going on, as if the photograph as such were a refuge today, only like Deckard in *Blade Runner* at the piano arrayed with photos that testify not to his obliterated personal life or memory but a history of the photograph itself before any memory.

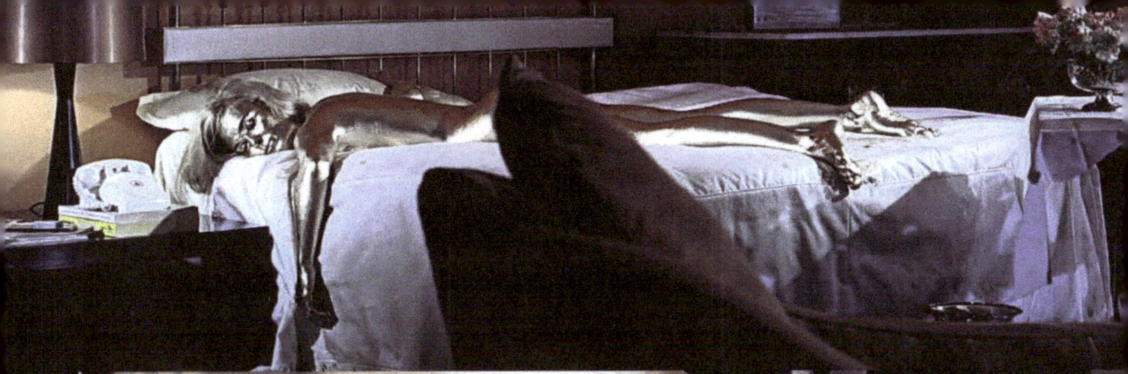

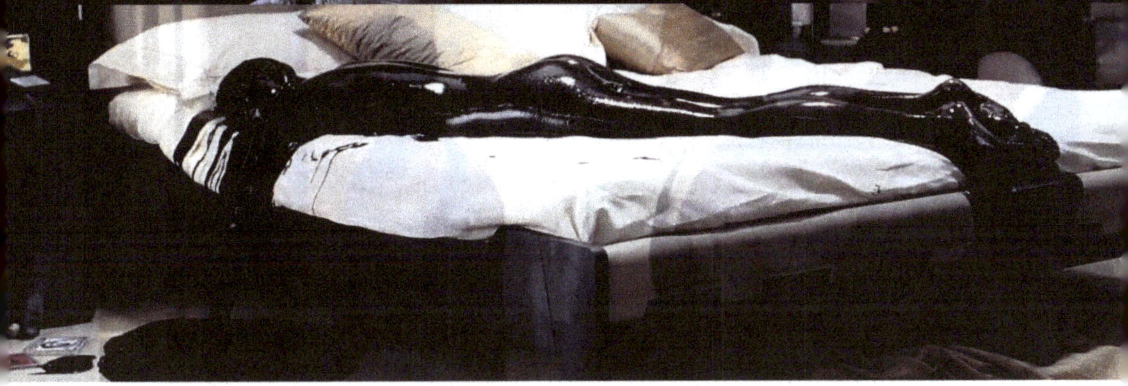

Now – I have been circling something unrepresentable that lies before and informs the frame, the "visible" as such. My first thought on invisibility and photography in the "era of climate change" made a wrong turn. I first contemplated addressing the new micro-photographic technologies, what discloses the invisible of the nano-world – in this case, two cute, new, micro-insectal forms never before viewed, catalogued or captured by our photo-telecratic imperium. They are astonishing, but they look violated, caught out by new

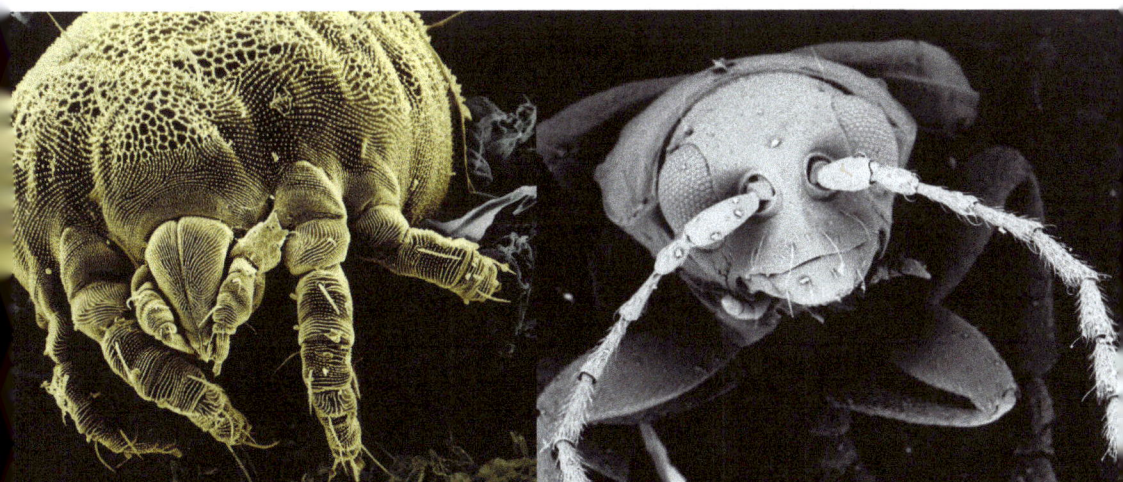

tele-surveillance – like the NSA activating your laptop cam. The next day I noticed in some news blogs that celluloid archives were under assault from micro-organisms, and I couldn't not make the link that this was a counter-assault, a militarization against forcing them into our organism death camps of visibility and use. I will relay the genesis of this essay topic, however, which required an external prod.

An e-mail arrived from Gabriela Nouzeilles – an e-mail, and I trace the genesis of this piece, include it here, since it spawned what is to follow. I will call this mode of tracing a "non-figure" of oil and itinerant photography (through cinema, in fact) a sort of "petro-telepathy," a writing by ooze and leaks. Gabriela was, accordingly, my muse here. And the trickle that seeps between frames, names and historial moments will lead through Robert Capa (icon of itinerant photographers), Sebastião Salgado, of course "oil," and then Hitchcock, cinema, even Rio itself, and dump us, finally, in the disaster of the Gulf oil "spill," the "worst" American ecological disaster, so called, an event that cannot (even now) be marked by one time or moment, however withdrawn again from the media gaze. What we call oil partakes, let's recall, of very different temporalities and geomorphic agencies;

> Hi Tom,
> Capra sounds good and I love the catstrophic promise in your title. I found these two photographs by Sebastiao Salgado that might be of interest. Salgado's approach is always problematic, but it invites reflection.
>
> https://www.taschen.com/en/books/photography/05314sebastiao-salgado-kuwait-a-desert-on-fire/
>
> I hope you have recovered from your Asian cold. I had a great time talking to you the other night. Un abrazo G

derived from dinosaurs and algae, it returns when it breaks surface for the camera.

The astonishing photograph by Salgado seems to index something other than reportage on the burning Kuwait oil fields--and in ways it spawned the meditation I will offer, one winding through oil, photography, cinema, and the enframed trap of today's waning petro-human. I will play with the term petrolepathy to name the mutating algorithm that runs through these episodes. That photograph is of a totalized earth covered in a skin of oil, as a laborer ("bronzed") tries to work a well bit. That cylindrical and static well bit, itself as if thrust into an earth, also hovers at the center--like the monolith opening Kubrick's film 2001, unaccountable amidst the worshipful ape-humans. Salgado's lens, too, appears transfixed, as if photography itself were called to identify with and be interrogated in turn by what is framed before and by it. (Note: [the image can be found on p. 9 of the Taschen site.](#))

III

It is here that the narrative bifurcates, and then again.

Gabriela's note encourages me to think about Robert Capa, icon of war photographers and itinerant photographers, and then references astonishing shots by Sebastião Salgado of the Kuwaiti oil disaster Saddam triggered as a Parthian shot. Two gifts here: the first is that she spells "Capa" with an "r" as "Capra," which reminds us that in fact "Capa" took his own name from the Hollywood director, Frank Capra (deleting the "r," the letter of repetition, and turning the name into, well, the figure of a head pure and simple), even as the name "Robert" was derived from an actor (Robert Taylor – perhaps you've seen the old *Ivanhoe* movie?). From the start, that is, Endre Friedmann, Hungarian Jew, invented a persona that accompanied the myriad shots of extreme photography, war moments, utterly embedded in the real and "life" at such moments, utterly at risk – and, in the end, taken out by a landmine in Southeast Asia (unlike his then young wife, Gerda, who would be merely crushed by a tank in Spain). So there is a point here that is never remarked perhaps: that "Capa" as the itinerant photographer par excellence, whose captured moments of "life" would be fearless and unprecedented, saw himself, nonetheless, as a Hollywood creation or effect, as embedded in a broader cinematic culture or hegemony in which "life" itself would be cinematic in advance of itself perhaps. Even his memoirs (*Slightly Out of Focus*) would be written to be made into a film (never pursued). More: his "real" would mark a claim of usurpation as well – as if jerking the whole Hollywood imperial riches back into the mud, the earth, the life and death instant to… well, I'll return to this, because the pictures by Salgado are imperative and, as Gaby notes, give rise to "reflection." That "reflection," however, echoes in Salgado's own comment on the first shot, in which he (Salgado) is fixated on how the one worker covered in oil seems arrested in time, become a "bronze statue."

There are three key elements to extract in this image. It itself is in one sense a "shot" of and at the so-called Anthropocene itself. In another sense, it performs a specular hiatus on "photography's" own enframing by this scene. First, the earth and sky appear totally suffused and covered with the renegade oil spray; second, the sole human as if frozen in a support position, labors all but hopelessly to repair or service or extinguish or maintain the main yet absent subject ("oil"); and third, the sheer (and non-humn) technic represented by the metal cylinder thrust into an earth which, like Wallace Stevens' jar in "Anecdote of a Jar," appears to order a world and horizons about it inertly. Sheer technics, to which the laboring "human," appears dedicated if not subservient.

What you see in this shot, overwriting Salgado, is a mise en abyme – photography arrested by its immersion in oil, the visible inverted black for white, the "bronze statue" outbidding the premise of photographic arrest, the drill an alien prosthesis transforming the earth in the frame, a sheer technics about which all "visibility" is reformed. The mise en abyme of photography itself, unaware, transfixed before its premise and creator – the specular relation of photography (freezing time like "statues") before the totalization of its source, oil, white and black exchanging places, the human forms set against a totalized horizon of what we will call, looking ahead, the link between the Anthropocene era, the photographic era of man, and oil – or the "light" and obliteration of the three. It is shocking, blinding, this reflection – and I move on for now. One is, at once, beyond any Benjaminian shock, mimed in the companion photo as a fireball – where "oil" combusts into fire, light and toxic clouds. "Reflection," indeed. Humans servicing a re-allegorization of an earth indexed to a photographesis in which the visible's apparatus is sunlessly inverted, and so on. And that's why the leakage I seek turns to the letter "r" – to "Capra" become "Capa" or restored to "Cap(r)a," to a pre-cinematization of the itinerant photograph or photographer I alluded to. And it is the precursor Capa, not Salgado, who strangely enough returns us briefly to "Rio," to which all roads lead. And, rather dangerously, to the one director who could read what is going on and respond with a devastating counter-stroke or (almost) slap-down – that is, to Hitchcock.

Capa, you know, wanted to displace Hollywood in effect, derail the cinematic spell of tawdry allegories by inserting the "real" as a take-back (assuming, perhaps, the relation of cinema to war itself – on which Tarantino's *Inglourious Basterds* has recently given a weird mock-up: the Second World War as a contest between Berlin's UFA

studios and Hollywood for world cinematic domination, and so on). To do this, he needed a trophy, and took, really reduced to bondage, Ingrid Bergman. After a bit (jealous husband stuff), he could only see her on the set of *Notorious* – the Hitchcock film that, set in Rio, turned on a Nazi plot to acquire nuclear fission, radiation bombs, even as the title mimed Bergman's "tramp" reputation, inculcating by proxy the affair with Capa (another fold). (In this sense, our double return to Rio here, marked in and as *"NotoRIOus"* itself, around the birth of the atomic bomb, usurped into Hitchcock's plot as a figure of cinema itself, marks "Rio" as the city of cin-animation and the screen itself, the "reel" or site of this exchange.) The duel was on. Perhaps it is not irrelevant that Hitchcock would recall, or profess to, that Bergman passionately propositioned him during this time (which he, unlike Capa, deferred). And Bergman, who, you see now, was a sort of go-between for photographic and cinematic power; a mere relay perhaps, like *Vertigo*'s Carlotta, would complain to Hitchcock that Capa found excuses to not marry her – that she couldn't travel with him to war zones, that he couldn't get the assignments he wanted, that he couldn't stay in Hollywood despite the drive to usurp it. All of which (you will recognize) Hitchcock took up as the derisive premise of the lame – immobilized and visually impotent – photo-journalist as James Stewart in *Rear Window*. Now again, we can't stay here long, we are only tourists today, but this slap-down by cinema as it swarms and disables the pretense of travel, of the claim of photographic power, is indexed to oil in the virtual deconstruction he, Hitchcock, gives to photography opening the film. His camera swarms and prowls (like a disembodied

The Viscosity of the Image

black cat) around the sleeping and immobilized Stewart/Capa who imagines himself immersed in travel and the real world itself (or so dreams).

The series of photos tells a story of the techno-genesis of photographs – the emergence over a crash (race car) in which speed converts into a catastrophic halt with a circular logic (torn wheel) flying at the eye or head (Capa), shooting (wrecked camera, immobilized Stewart dreaming). The concluding negative of a generic blond emerges as the generic circulation of a fashion mag cover, tossed in a pile, a grimacing mock-up of a Grace Kelly that takes over Bergman's role as a photo (or statue) come to "life," sexually anathema. And in between a sequence of crashes, explosions, mushroom clouds silhouetting human shadow forms – say, against a giant oil fireball, and then a discreet mushroom cloud, citing again (as per *Notorious*) a specular link with atomic blasts, the atomization of the world into particulate matter, irradiated, that cinema practices in fact and by extension. (Hitchcock portrayed the atom bomb, by the way, before it was known or released or used, causing the FBI to visit him and ask how he knew – considering censorship – not getting the logic of the MacGuffin.) Anyway, cinema here totally swarms and cripples and slaps down the usurpative Capa and an "itinerant photography" displacing cinematic logics; it embeds the latter in a cinematic aside, marks it, brackets its dreamed power in isolation and its claims to real-world transport, travel or "life."

Capa got close – he got to be "Capa," not "Capra"; and he got Bergman, from under Hitchcock's nose, but then he bolted, and drifted back to step on a landmine in Asia. Hitchcock pounces and, so to speak, "de-Capatates" photography. I'll return to a footnote on this later, in which, under Hitchcock's own hand, he reverses this relationship for a moment, inverts it, and captures what may be called the "era

of cinema" in a snapshot on the set of *Psycho* – a still shot, we might say, of the cin-anthropocene era tout court…. Bergman is discarded as the oil-draped woman-body, go-between less for the two males than a discreet war between the two media, intertwined, inextricable, fratricidal, framed by and before a petrol-telepathic catastrophe that is channeled, and interminable, leaking into or framing this room today. Dissolve. To seek a face for "oil," one turns to where it makes itself seen. And to where, momentarily, cameras cannot get enough of it since, *avant la lettre*, it is perpetually ghost-like. But to ask of oil how it frames the era of cinema is to flip the way in which we encounter the photographic image – it is to opt out of a system of image circulation, much as "Occupy" opted out of a totalized, economic-ecological übernetwork that had no apparent outside to its vortices (easily tracked and disoccupied). With this in mind, it is para-political to suggest that the eye occupy the position of the invisible. Perhaps one must opt, before an entropic, "Anthropocene" enclosure, for the eye to "occupy oil"? Negarastani writes a visionary solicitation of this perspective, which he calls "globjectivity," but it turns out that has always circulated discreetly in the reverse side of any photographic interrogation.

IV

One is routinely tempted to think of oil as at the invisible core of hyper-industrial transformation. It suggests a carbonic link to writing, a liquid mass preceding or dissolving letteration. At a supposed interface of the organic and the inorganic, the black storage of solar technics undergirds the hyper-industrial accelerations, the autophagy

of "life" on its own waste. Necessarily, one requires an "event" where it breaks surface, claims visibility. The BP Gulf Oil eruption ("spill" is for a teacup) offered a plethora of public photos vying to witness, to make visible – and this, despite a corporate-governmental-media curtain that lowered immediately to police the eruption. At the "time" – but what is the time of such an event, certainly not over "now"? – speculation ran amok: would the oil plume (a nice word not unrelated to "plumage") reach the north Atlantic, create a sheen on the surface, collapse proto-plankton and oceanic food chains, leave us a sea of jellyfish to eat, and so on? Cinematic imaginaries.

But here there is no authorial composition, no signature by which to read. There are aerial or testimonial photos. From the other side of the equation, the side of "oil," patterns and genres emerge that are self-selected by their recurrence. I will just scan a type of "shots," which almost form idioms or patterns.

The invitation announced in David W. Dunlap's layout of "Putting a Face on the Gulf Leak" is suggestive – despite the suppressed nomenclature of "leaks." To draw us in, to bond our eye, oil needs a middle space of half-personification or forms. An iconic, oil-covered bird. We get it. It needs a face for us to invest, and a claim to pathos – the elicited response of a doomed, oil-covered pelican. We are asked to imagine the eyes looking out, thinking its zombie status, a still-living tomb. We humans are given faces (more or less: do animals have, in that sense, face?), and a shape that the oil drapes itself over. We are prepped to enter a curious, half-mourning zone of the photograph, either guilty or relieved that it is another (species). But the solicitation ("putting a face on... ?") can as well be a discreet mirror – it is not, really, the pelicans that are so immersed and defined head to toe, entrapped and defined thus, as Salgado marks and the Bond film tropes lusciously. In this contract of the in-between, one covers a broader defacement, so one must ask how to read oil here: what precedes form or letteration, how to extract the perspective of what is not a subject and not an object. One enters a counter-zone in which the late "Anthropocene" or hyper-industrial present is as if interrupted by what is indifferent to it. That is, what is like waste, or shit, a black hole that gives light, an undead storage site of sunlight, technics, a hyper-matter devoid of "materiality."

V

The notorious BP Gulf oil "spill," which Hollywood long ago built buoys around to narratively contain, was (and continues to be) about visibility – hence the photographic testimonials and genres at the time obscure a certain grammar. University research and coastal police were bought up to suppress details, much as the dispersant that sank surface oil at the price of greater contaminants would delete it from the surface, while the gallery of soaked, poisoned or encased animals distracted from their being metaphors and stand-ins for the viewer. One contained temporalities as well (since the drift and implications continue to seep and mutate out of sight). So one can only break up a blizzard of photo-testimonial, apotropaic as information or data, in a rare instance of "oil" exteriorizing itself, coming to the lens massively, altering the makeup of the visual and visible, while the entire exvasion is channeled away from its uncontainable implications, participation in orchestration of climate collapse and technological overdrive, the blood accelerating AI inundation and climate dissolutions jointly. So, one is left reading "oil" as refracted through photographic images that – shaped by implicit genres (mud zombie creatures, vast discolorations, hands dripping) – suggest a grammar and dangerous zone of inmixing one might mark or open otherwise in reverse. This, keeping in mind that what we call "oil" does not quite exist as a thing (or even as the *hyperobject of hyperobjects*), is not encountered directly, is toxic to touch, yet performs the extraction of the residue of organic "life" past, the Nethercene, back into an atmosphere it had been removed from ("stored sunlight"). How does commercial photography, artful and not, engage or defer this breakout?

I here select narrative patterns that are formed or repeat across photographic reportage. So much so that a certain grammar without

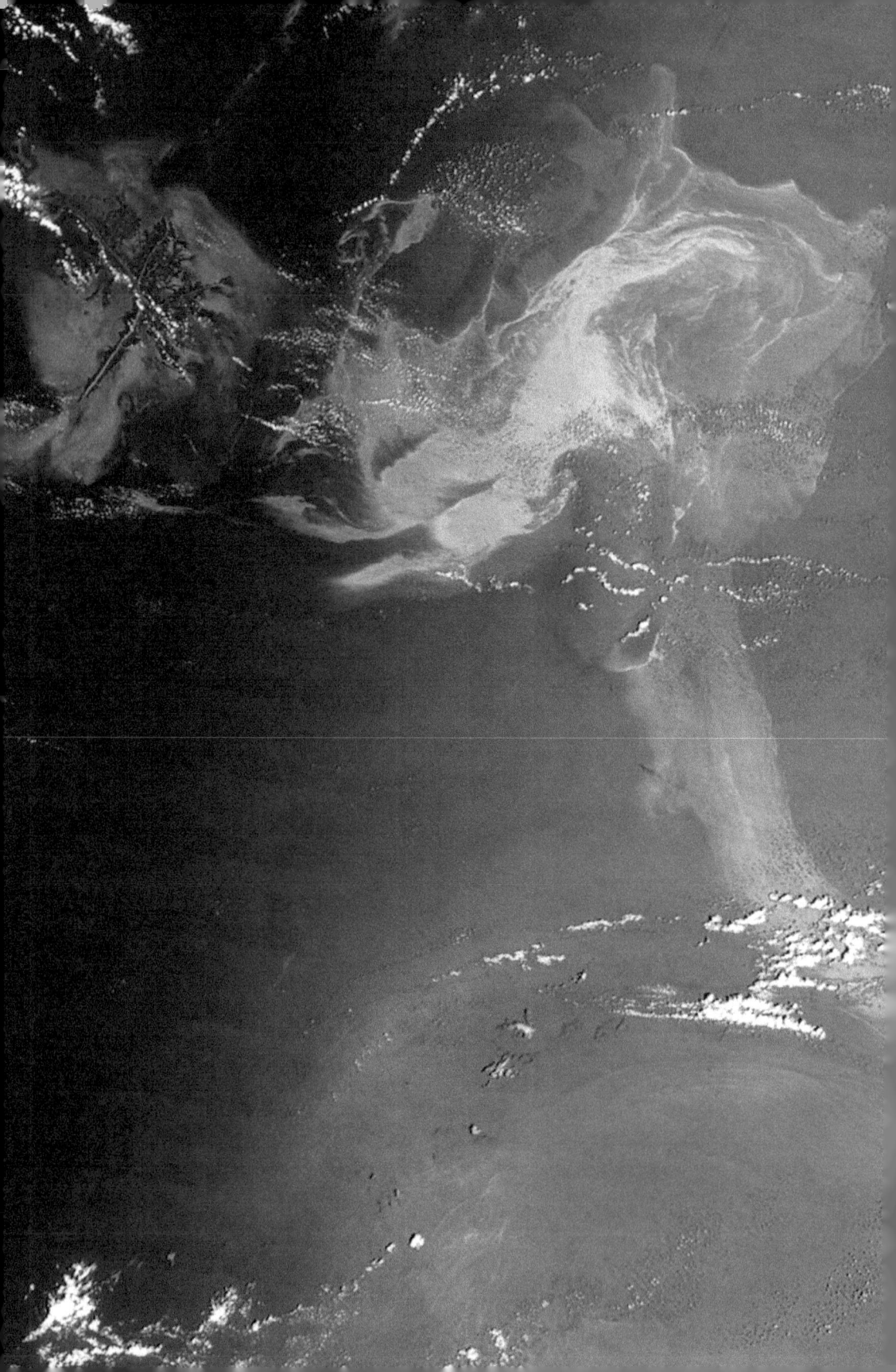

authorship emerges in the mnemonic negotiation – for what was to have been testimony. As clusters, an image-language of a non-subject, "oil," arises, diminishingly labeled as a "leak," or a "spill." Given its diverse modes of (in)visibility, it would be put out of sight again. What becomes apparent, and distracting, is something like the abject beauty and proto-aesthesis.

Surfacings. Aerial, the terms of oil's visibility is differentiation, what divides a surface from itself. One may speak in this sense of a "gulf" within oil, a gap that it cannot divorce itself from...

Attacked by oil dispersants to render "the spill" invisible from the surface, suspended beneath the surface to not be seen, it does not sink, but aggregates. And it holds, on the surface, a different light or sheen. This sheen, able to deflect light from the waters beneath, matches to an encasing threat below. Should sunlight not penetrate, the sheen or skin becomes totalized.

Between liquid and form, energy and matter, organic and inorganic.
Is this mere narration where there is, unlike the discreet combat of Capa and Hitchcock, no storyline precisely?

Engulfment, Plasticity. The aggregation shifts, darkens, mimes a corroding celluloid that feeds by withheld assimilation – drawing itself around a prey, flashing monstrous hues.

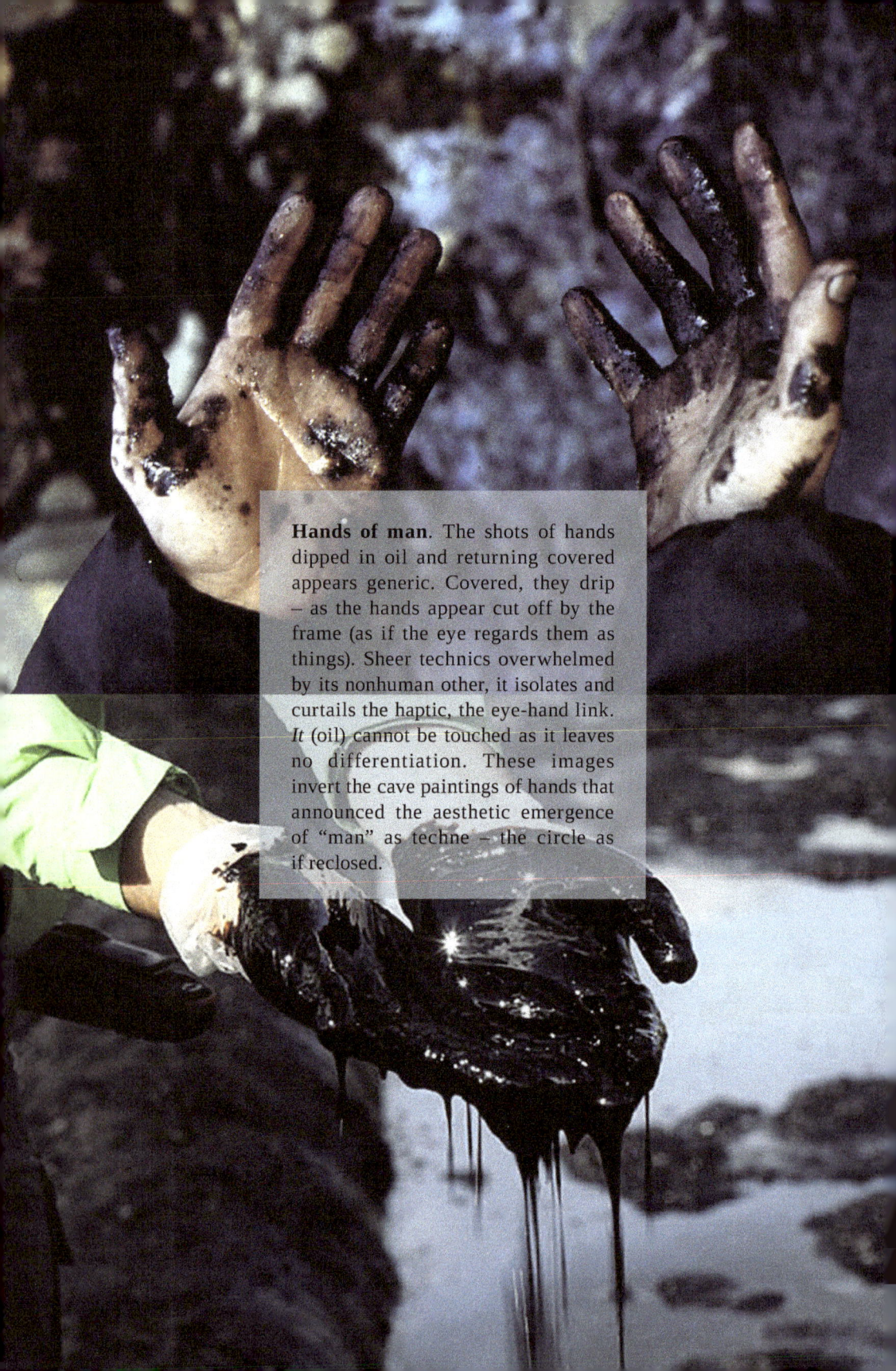

Hands of man. The shots of hands dipped in oil and returning covered appears generic. Covered, they drip – as the hands appear cut off by the frame (as if the eye regards them as things). Sheer technics overwhelmed by its nonhuman other, it isolates and curtails the haptic, the eye-hand link. *It* (oil) cannot be touched as it leaves no differentiation. These images invert the cave paintings of hands that announced the aesthetic emergence of "man" as techne – the circle as if reclosed.

Militarization, containment. Humanoids push back, would surround, defend. The phantom of containment is militarized — against what, derived from aeons of organic life forms, absorbs and is the trace of so-called species.

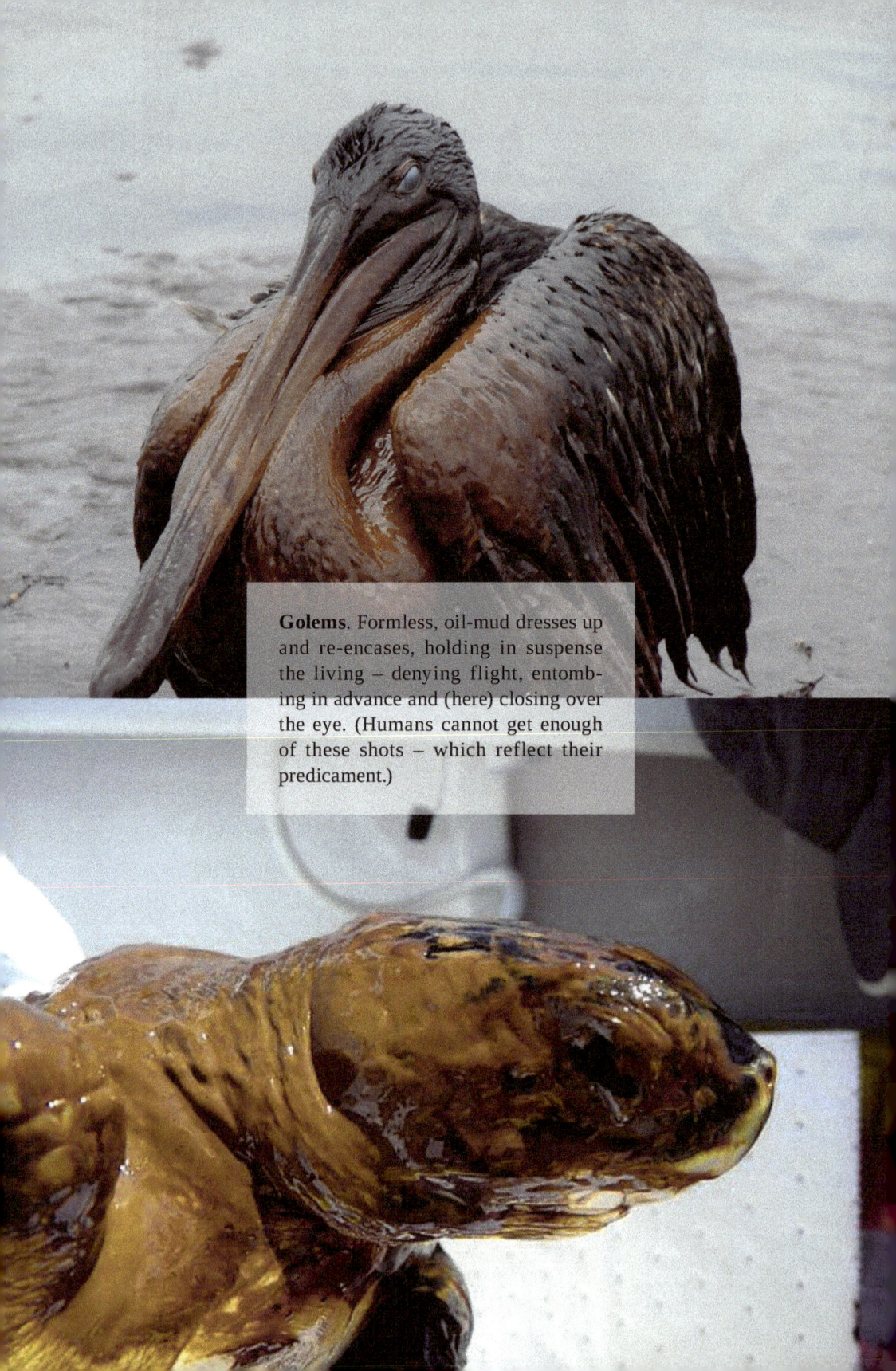

Golems. Formless, oil-mud dresses up and re-encases, holding in suspense the living — denying flight, entombing in advance and (here) closing over the eye. (Humans cannot get enough of these shots — which reflect their predicament.)

Oil at play... The splayed torsion of counter-twists conjures the phantom of abiotic oil (that is, a hypothetical blood of an earthcore purged of any organic decay or "life," simulating the fantasy of an oil pure of death and plentifully unavailable – what would conjure an oil at once tamed by organicist metaphor yet rendered un-obscene, paradoxically free of death...).

Here, it streaks, miming a sheer acceleration of speeds...

Oil in jouissance.... The pyrotechnic release and transformation into fire, heat, air, black cloud fury. Drawing the pygmy boats and water jets to encircle and suppress.

VI

But of course, this "putting a face on the oil leak" mirrors the viewer or creature in fact swathed in oil products – since the net effect is to draw our facing techniques out only to be claimed, like a mud-covered turtle. These routine gestures mime the moment of militarized containment above.

It is precisely this defacement, however, that cinema implies when it thinks the black light and atomized particulates – the mud – that it animates. This appears to have authorized Hitchcock's deconstruction of photographic techno-genesis when capturing, and emasculating, his Capa. It marks the awareness of a cin-anthropocene era that is destructively interdependent with the era of cinema that coincides with it. Humans are merely the agency by which oil arranges its mirrors, hanging back like the "do not forget" of Hamlet's ghost. I do not have time, of course, to trace the many links of this petrolepathic network entirely evident in Hitchcock – whose most literal moment is the close of *Psycho*, perhaps, when Norman's face dissolves into "mother's" skull and that nonexistent "mother" (she appears nowhere in the film) dissolves in turn into the bog or mud-swamp into which, oil like, cars too are absorbed, the very vehicle of transport. And we see it early, just to mark this, in the silent *The Manxman* when the judge, himself culpable for the child that the woman before him is being judged for having, dips his pen in an inkwell that dissolves into a black drowning pool which claims a suicide – the "law" itself written, sustained, illicitly, against this. And it is in *The Birds*, of course, where this cycle is hyperbolically on display, as these attacking animemes and wings, which shatter this black pool into myriad black points now in the sky that attack – and peck out eyes, considering the contract of the viewer to the screen itself a blindness

and trance. Descendants of the dinosaurs whose remnants run our machines, themselves sheer technics of the screen (machinal in turn), they bring an entire cycle to a head when attacking the gas station and releasing fire and light in a chaotic blitz in which hangs a swinging sign, "CAPITOL OIL" (we keep this term, "capitol," "capital," and "capo," or "Capa," in play). With a little manipulation, the entire cycle or closed circuit we pretended to trace, above, in "putting a face on the oil leak" is recapitulated in petto. A petrol-cin-anthropocene?

VII

I turn to an inversion of the *Rear Window* capture of photography by cinematic logics – the placement, and slap-down, of the upstart Capa's claim to be the living eye, or head, of cinematic logics in "life," as travel, tearing apart the corporate artifice of Hollywood by usurping it as "pretender." One might call it Capa's revenge or that of still photography. Hitchcock cannot resist chiasmic reversals, least of all against "himself." How, though, can a photographic shot encompass, fix, compose or display the era of cinema, its temporal logics and expanse and multiplicities and even future? What would a still photo of the cin-anthropocene era tout court, with all its marking systems and destroying backloops, even be?

One day on the set of *Psycho*, Hitchcock cannot resist doing just this. It is all there, by marking itself as a "set" – with Tony Perkins "as" Norman visibly standing on an "X" to position him. That is, the shot is of Perkins as "Norman" and the film *Psycho* itself – what, to borrow the facile language of Žižek, marks the "subject without subjectivity" of the cinematic era; neither human nor not human, he (Norman) is placed in a frame before a desert earthscape. The bog, it

The Viscosity of the Image   51

is implied, this mother that is not (think of Derrida's figure of *Khora* here, the non-site of all inscriptions that is not gendered, not phenomenal, not susceptible to touch since it programs mnemonics). The bog is off-screen but implied as what Norman-Perkins stands before (and poor Perkins, who is sacrificed into this moment, who can get no acting jobs since he is forever "Norman Bates" and ends his days, an early AIDS case, returning to the spot, cranking out *Psycho* 2, 3, 4, and 5... impossible to get out of this shot, since it inscribes and absorbs the "real" in its cinematized logics and the event). And he, man, is displaced from the center, abutting a giant dead tree – a ghost formation that is given a comic face by shadows. The sun is setting in the west, and the crotch of the "tree" displays a weird half-personification that takes over.

A horned totem, bull-like perhaps, half-human and half-animeme, monstrous, it conjures briefly, in the general index, a Minotaur within the labyrinth of the visible – what gores or consumes its visitor. But it changes again, the face that is only implied by light and shadow from an off-site sun is of a dead tree consonant with the eviscerated landscape, of which a tree, like a mushroom cloud, stands at the top. The "tree" converts into a sheer techne, not organic but thrust, like Salgado's drill-bit, into an earth and transforming it within the circulation of photographemic logics: like a single bar, a one with branches, or a letter, a gamma with cross bars (third letter of the ur-alphabetic, the non-origin of number in the scandalous invention of a "one" and a "three" to which Hitchcock links the advent of cinema and man in a fashion I cannot elaborate here – this "13," Hitchcock's birthdate (August 13, 1899), recurs in each film: think the house number of *The Lodger*, and so on). And it has a gap, an elongated circle or aperture in its non-trunk, a "zero" absorbing the Zarathustran logics of the circuit, the circle, the temporal backloop that cinematic time names.

Hitchcock switches roles. He turns to photograph, that is, shoot, cinema itself – in its totality as a moment, the era of man, the cin-anthropocene, set before an earth eviscerated of organic life, nature, trees. Hints of structures litter the bushes, derrick-like, extracting light from a sunless, black and white scape that is itself produced by and indexes the cinematic technics that partake of these eviscerations. Man as the era of cinema; "man" as Norman displaced by the technics of letterations and semiosis, the Minotaur non-face giving way to the circuits of return and the zero logic of this order of the photographic event. A snapshot looking forward and back, of cinema, to be filed in some time capsule and to be read as if by others later. Capa's revenge, or Hitchcock taking up the now headless lens to shoot to kill, capture, distill or fix in a permanent non-time its entire circuit – forward and aft. What emerges is Hitchcock's shot, if you like, which is nugatory in itself, of the cin-anthropocene in its totality and arc. Norman's bog, outside the frame and ostensibly where the camera shoots from, is the La Brea tarpit of spectral histories of the cinematic petro-image, light-writing (photography).

This returns us not only to the Salgado photograph we began with, which has a certain resonance in the accord between the sheer violence and centrality of the drill piece, the radical prosthetic of this gash and theft which all is referenced to, but to the survivors – outside the loop of this temporal parenthesis, the "time" of the photograph and accelerated resource depletion, mass extinction events, and the delivery of the stored carbon of earth's aeonic dead into the light. Salgado's upright drill, as if thrust into an earth yet not of it; Norman's dead or ghost tree; Kubrick's monolith opening to *2001* appearing already rammed into an earth organizing ape-anthropoi communities. That is, toward the micro-organisms who evade what is a fairly quick episode from the point of view of geological time, or oil itself. It is a rather pointless exercise to give a face or name to these fellows, to repeat the last gesture of anthropomorphism and incorporation we are rather pathetically programmed to do – even as an "insect Armageddon" is reportedly underway.

To withdraw from a totalized system that appears to be without exit itself, say, parallels the viral, if passive, insurrection of "Occupy" movements for the brief daylight they were permitted. The insurrection would "occupy" the source of an invisible totalization gone awry or captured. With this parallel in mind, to seek or occupy the invisible premises of "visibility" is a para-political gesture – a displacement of a system of circulation itself. This would be the imperative, so to speak, to locate within the eye the invisible factors constituting what appears a suicidal blind (the era of eco-catastrophe, say). The eye might better inhabit the technics that give rise to the visible; the eye might practice occupying what does not, strictly or as a word, exist – oil. Particularly when the latter's logic paints us as energy-besotted, oil-suffused even as we speculate on and overshoot its "end," the end of the oil era, the dawn of "renewables," of sunlight, of wind.

In tracking the entanglements between photography, cinema and oil, we navigate a complex web of visibility and invisibility that flows through mutating switchboards and spells. From Capa's war photography to Hitchcock's cinematic prowess, from the Gulf oil "spill's" stark imagery to the subtle omnipresence of oil in our daily lives, we are confronted with the paradox of our petro-culture: ubiquitous yet unseen, foundational yet contraband, one becomes alert to modes of petrolepathy. The grammar of oil's visibility in photography – whether through engulfed wildlife, militarized containment efforts or abstract surface patterns – seems apotropaic. As we stand outside the cinanthropocene, captured in Hitchcock's meta-photographic moment on the set of *Psycho*, we are challenged to locate our place within this increasingly soaked and flightless petro-telepathic network.

Ink, Poison, War

Hypothesis: Avital Ronell's early rogue writing, here Crack Wars, generated a unique trope, "the catastrophe of the liquid oozing," she calls it, that runs (or melts) through critical and literary writing – leading toward a war site of liquified inscriptions

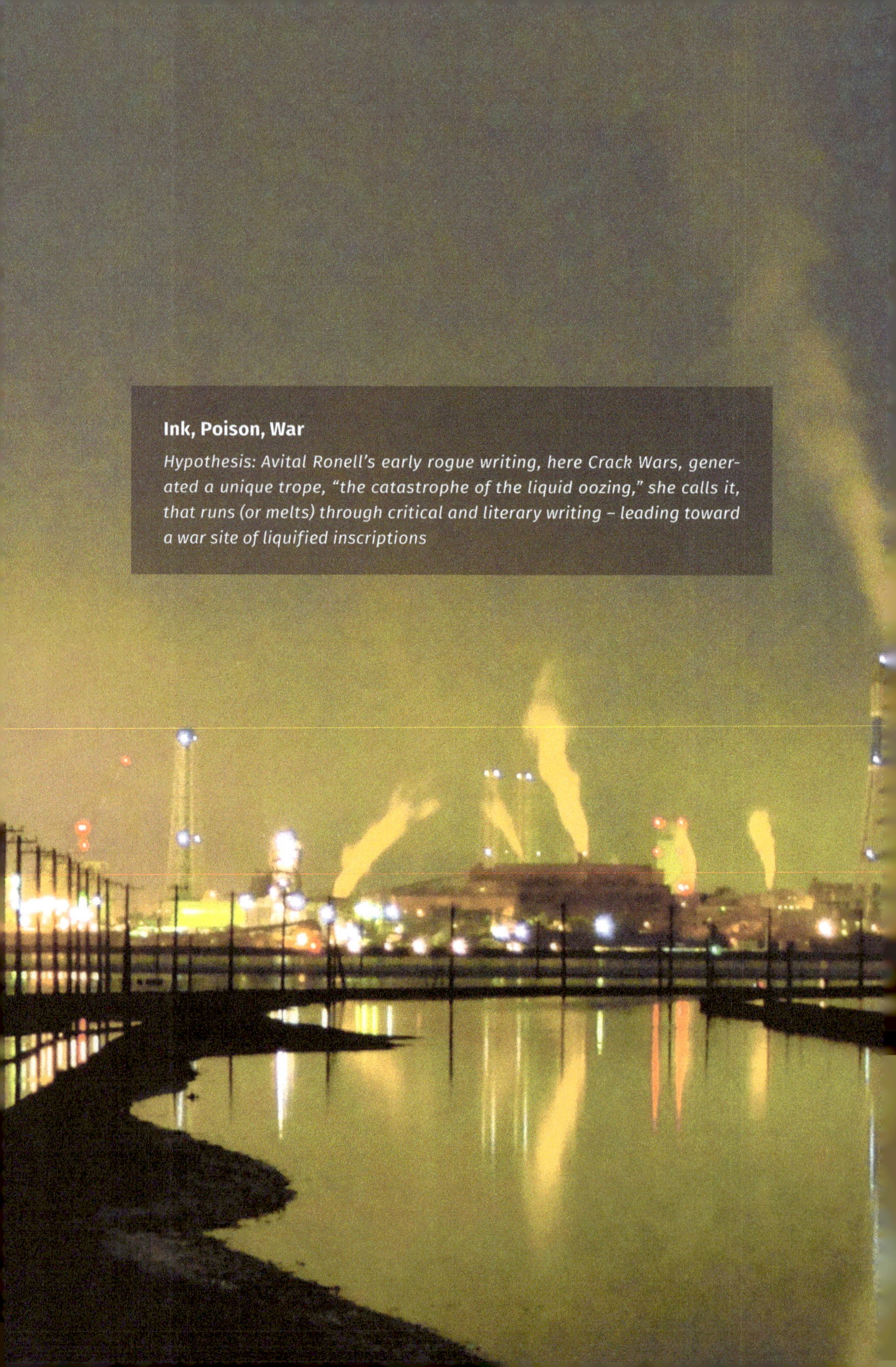

2   Roaming Discharges: Avital Ronell's "Catastrophe of the Liquid Oozing"

> Abstinence – cold turkey – opens the medusoid rift.
> —Avital Ronell, *Crack Wars*

> Coming from her, originating in her, it is nonetheless a foreign body, ever replacing the newly born body. Catastrophe of the liquid oozing
> —Avital Ronell, *Crack Wars*

This essay excavates a particular turn in the writing of *Crack Wars* that opens onto the problematic of the pre-figural, of ink and of liquefaction that might appear suspect, today, by having no obvious place to gather. Where does the petrolepathic circuits making cameos in the industrial image – an inveterate flasher – drift and flow in the figural "house" of writing and letteration?

One might approach Avital Ronell as a political scientist of memory transmission whose performative forays – at plague centers within a *grand mal d'archive* – negotiate a different relation to the catastrophic. There are vapors one encounters in this prose, drugs without names of the sort that concern Ronell, and one enters corridors within her syntax that beckon, or unravel, into accelerating passages. One could assemble a file of these, trip over their accumulation and vanishings, get off in uncharted spots at which the body writes under other names, under the radar of swooping policial cropdusters. Peel back the analytic riffs and one finds oneself, forgetting these opiates, at the revocation of imagined histories: legatees are threaded like beads, epistemopolitical machines are exposed, and one is delivered, if one is, to sites one cannot get back from. One finds oneself – no, I do, who find sanctuary in a certain tonality – before an array of portals that whisper again and again one word: "war." Ronell goes out for drives to test this perimeter. There are crack wars and drug wars, the tropes of Desert Gulf, war as illness: one hears the "tensions rumbling through the novel derive from a secret war against artificial, pathogenic and foreign invasions" (*Crack Wars* 115);[11] "the poetic and war efforts appear often to interlap" (*Stupidity* 5); the "warrior impulse" (110), and illness is itself "war" (186).[12] It is a word that recurs like a stamp, suggesting a "civil war" internal to the archive and contemporaneity, of which *Crack Wars* is an unlikely cipher and *Stupidity* a strategy of transition. Ronell roams the modern legacyscape, inspecting catastrophes of cultural transmission and their cognitive politics the way a ghost returns to a remembered site – from after "the transvaluating machine was left running" (*Crack Wars* 69).

I will examine the slow drip of this word, "war," but do so by way of the least auspicious trace among Ronell's preoccupations – the black ooze, which haunts the center of *Crack Wars*, coming from the mouth of Emma Bovary's corpse. Site of speech and ingestion, the running fluid that dissolves interiors in literature's famous corpse is like the acid-blood of Ridley Scott's alien, manifesting prefigural properties: ink-like, it seems to precede not just tropes but letteration itself – as if inscriptions were being liquefied.

What I would like to explore is a minor figure in her text, a bile she finds coming from the mouth of Emma Bovary and locates at the center of *Crack Wars* – as if, threading the pharmocopoetics of the "literary" and programmed culture, this figure had something to do with war itself. I will interpret this as a sort of prefigural agency Ronell encounters in secret places – a leakage that evacuates the remaining debris of old models of interiority. I will suggest it also condenses the figure of "anteriority" itself to a mercurial and corrosive ink, a site where inscriptions seem to have melted back. And I will say that, in its way, this figure, into which traditions may be dissolved as into an allohuman black hole within memory networks, is connected to a site linked to what Derrida calls "khora," a non-site where the preinscriptions from which reading models and meaning systems derive are set or effaced. Because of this link, the prefigural agency Ronell taps into anticipates coming wars of reinscription, which seem palpable on numerous epistemological fronts "today." The war, then, will be in and over archival programs and memory regimes that return, like oil, to the site of "catastrophe of the liquid oozing."

Why can, or must, one return to war in Ronell – even if that is called by other names, like "test drive"? Do others not know this war is going down? What has it to do with wars of transmission and legacies, of the archive and anteriority?

Ronell is the performative (an)archivist of a certain going under – the jacket of *Stupidity* speaks of "the fading of cognitive empires." Of course, she will be forced to migrate along the filaments of metaphor: from telephonics and switchboards (s)he is forced from the *terra* itself – which has been dissolved into circuits and cognitive mafias – into orbits, of which the iconic text or "name" thought secure operates as near space-junk and satellite ("Kant," "Wordsworth"). A "satellite" is turned toward the earth, teleporting memory or cognitive clichés, but its outer side faces constellations without anthropomorphic echo. Ronell's appeal to satellites mimics the war fantasies of a coming American panopticon, sensors of the "compressed kill chain" – a "Minority Report" scenario of preemption, of miscalculated temporal loops. A hypothesis: the war at issue is over inscriptions from which perceptual programs and legibility are generated, and not this or that territorial or colonial skit – drug wars or desert storms, wars on terror or involving academic self-mutilations (for the greatest drug is the ordering of certain cognitive rituals here, certain blinds). I will risk an obscure remark that may betray my stupidity: she wanders into a preparatory space of auto-sacrifice without a cut, and without a call, in which anteriority as "recognized" is also liquefied. She has wandered beside a non-site she calls at times "ex-scription." The epistemo-political war that is at issue may be more decisive if invisible today than any mere world war, since it condenses the historial labyrinth of the archive of the book to the point of a non-question.

Yet having said this, I am interested in something very small and irreducible that drifts through her tropology – a black liquid, a poison of sheer anteriority, a sort of vomit or voiding of interiors, the home, the family even. As the (an)archivist of a certain going under, Ronell seems unable to track the inner history of ideational forces, mini-genealogies, without exposing it to something like an encapsulating back-glance designed to close out a repetition cycle or mark where it is arrested and bypassed.

Ronell links what she calls "crack wars" to this figure at the omphalous of a seeming canon – the "novel," modernity, Flaubert. That is, to the black ooze coming from Emma's mouth in what one calls "death," which triggers a reverse temporal flow. The "catastrophe of the liquid oozing" marks time, is para-menstrual, drains:

> it all comes from the issue of her body, the sudden spill of liquid, the way she's stained and shredded by anguish. Coming from her, originating in her, it is nonetheless a foreign body, ever replacing the newly born body. Catastrophe of the liquid oozing. (*Crack Wars* 110)

The literary hit, "woman," the pharmacopia itself implodes here to an inky exteriority, a liquid black hole, a trickle of mnemonic transmission fluid. It alters the aesthetic montage we call "body": "Now this drainage which in itself produces nothing – there is no transfer of energy or funds – will terminate only when the cash flow gushes out of her mouth at the scene of her suicide. This is when the concept itself of currency becomes assimilated to her circulatory systems" (111). Even "coffee, a dissolute pleasure, brings up the haunted image of black liquid" (142). And: "(s)he does not manage to eliminate any particular force or figure, though it is said that ink flows from her mouth" (95). Ink-like, the ooze here vomited from Emma, which resists figuration, is not unrelated to the fluids of Norman Bates's oil-like bog – allied to what that film depicts as a "mother" without place or locus and to what the preceding film quips about as a sort of "alphabet soup."

Ronell arrests not just Emma Bovary, but a macabre instant of transmission that dissolves then discharges the "book's" intestines. What catapults the repetitions of an addiction – that is, at once chemical, semantic, referential and temporal addictions, even those supposedly birthing "modernism" (or its feint) – to a sort of hemorrhagic fever where the borders of cells and organs dissolve?

This gear shift metastasizes in Ronell's reading of Dostoyevsky in *Stupidity*, when Flaubert's book turns up in the pocket the "idiot" Myshkin. In this "coupling" between the two novelists, something passes: "The coupling with the other work seals the suicide pact" (*Stupidity* 223). Ronell reads this empocketing: "Depositing the book near or on his body... , Myshkin proffers his body as an impossible pre-scription, overwritten, as it were, and conscripted by a drive that comes from elsewhere. (There is no prescription for what he has.) Parasided and harassed, he, like anybody, finds himself borrowed and read as the map of expensive hospitality, an inscription pad where everyone crashes... Sealing and concealing the book, Myshkin signs in and under the name of the other, binding himself irrevocably to this power that comes from elsewhere" (241-42). Ronell here isolates a unique site, little accessed in today's criticism, an Odradekian petratrope amid her exorbitant traffic and eavesdroppings on teletechnic switchboards and the inner histories of dead-enders, suicides, ecstates, addicts and those beyond mourning. We witness in Ronell's detective work on cultural mnemonics the encounter with a khoratic agent, where inscriptions have dissolved into a prefigural soup in and from which vehicles of transport are dumped or retrieved. The black fluid dissolves script or letteral shapes to some preoriginary stuff. Ronell draws near to this non-site to wrest from it other times and lines of force, alternative time-spaces or historial backloops to which the ones we call "tradition" represent façades and relapses (to use de Man's cancer-inflected trope, to which Ronell, with clear antagonism, draws close. One could call de Man a "transmorphic repressed" of Ronell, but (s)he names this obliquely, virtually dedicates *Stupidity* to that, if with a palpable nausea, as if accessing this anti-poison to Derridean sweets, the protective feint into the staging of "ethical" pointers and aporia that would draw off the "Derridean" moths yearning for a good to claim as much as Derrida required, after the de Man debacle, to weave and generate that as a protection against the trolling hermeneutic police aiming to add his scalp to the list of the proscribed (de Man, Heidegger) – at the expense, it seems, of being able to take in the disclosure of the gathering extinction logics that would, at the time, have abrupted that ethical weave.

The black oozing – what has this ink-like stuff, not quite vomit, to do with a blinding "white" of the page, of (a) crack, or the war that is at once, today, invisible and so totalized it seems without temporal or geographical horizon, an epistemographic trance or eddy like the façade of a global war on terror, like white noise? This crack war has been there a long time. Already, it was that totalized war that Benjamin responded to in the *Theses* when he spoke of the "enemy" as historicism or a certain media programming of perception (and not, that is, the mere "world" wars of late colonial fascism in its proto-technic, genocidal convulsion). That is, what is totalized in the contemporary fever of mimetic and mediacratic programs, commodified reference, memory regimes. What war, if what "was" itself sheer anteriority, leaks or recongeals – if anteriority, like this black discharge, offers itself, as it is shown in *Crack Wars*, as the cipher of the phantasm of a "modernity" that, today, seems so anesthetized, so drugged? Ronell chooses the back trails of the literary to operate, in mock-guerilla fashion. What she liquefies bursts the cell walls, runs under portals, eludes surveillance.

There is a question about the "Americanness" of Ronell's project. Ronell writes forgetfully from after the "transvaluating machine" has been left in default. She assumes a teletechnic switch that displaces and consumes the protocols of the book, disclosing networks that precede it. (It is typical of Ronell's "books" that she marks, pleasures in and distances the latter's production as commodified object and memory bomb.) Ronell leaves a marker of this American or totalizing effect in the veil put over the body of a simulant, Emma, a disinfectant-effect. That veil is designed to lessen the corpse's smell, to render slow-motion the catastrophe of the oozing liquid from migrating virally. When Ronell loops back to *Crack Wars* through the dossier of Dostoyevsky's *The Idiot* – through Prince Myshkin, who carries in his pocket a copy of *Madame Bovary* at his collapse – she maps a contamination in the tele-networks of the literary, a wholly other model of "literary history" to anything available in the hermeneutic pharmacy. Ronell chooses to focus not on the female corpse of Natasya, an Emma avatar, but on a special American "cloth" connected to smell, to interiors: "This body, now reduced to the smell of preservatives, is covered by a medicalized trace called America: 'Do you notice the smell?' …

'I covered her with American cloth – good American cloth...'" The corpse can be lightly concealed by the smell of "Zhdanor's disinfectant." The black ooze is active, like a viral agent, leaping from textual bodice to bodice, accelerating its reclamation of whatever pretended to be extending temporalities by this linkage. There is *no sanctuary*, no body it does not burn through or reclaim in a back-wheel of temporal extensions and prefigural premises. Ronell is unaware, at this point in time, and may remain so, of the black drip of extinction events surrounding her, the trickle through the inverse poetics of the era of climate chaos, and above all oil, that informs a liquid black that precedes "light" or color. Melancholic, but that, while refusing reference itself, fossil-ridden, turns out, in secret, to inform and inhabit what is called "light" (or sun) as a technic effect itself, the effect of interval, nuclear combustion, what is called fire. There is no sanctuary, then, from whatever pours from Emma's lips – whatever used her literary corpse as a vehicle of detransmission. In another satellite, what we will call "Faulkner," operating within a totalizing Americanist cloth or canon, the figure of Emma and her black fluid is again pocketed, although with accelerated and jamming results – moreover, *it* will coalesce as a personified character, a figure, a gangster, an emissary of graphic animation and pop culture.

Once "literature" is relieved of its institutional definition, as Ronell's readings presume and perform, it discloses itself as a teletechnic of mnemonic networks. This is why one may pretend to speak today of climate chaos as still having a literary structure, as an effect of writing practices it cannot outrun but only accelerate in digital vortices. These networks do not enforce mimetic and identificatory, archival and interiorist reading practices or knowledges – they may seem abject, "literary," stupid, perhaps impotent and frenzied before a certain screen they cannot traverse. These are, as Benjamin says of cinema, de-auratic, which is to say before personification kicks in with its trances and transferences. These are no longer tropological primarily – which does not make them without some sort of direction, or at least an irreversible status. They scan transactions before which the humanistic models, neo-Enlightenment and hermeneutic programs, appear simply in evaporation – dispersed across other signifying-acts machines ("A mere copier and data bank attached invisibly to a larger apparatus, I am programmed to situate the problem and respond to its call.... I am going to have it scan the entirety of the argument as it sifts and sorts, putting the information into a new order" (*Stupidity* 280)). And "literature" was always doing this, steeped in the drug trade of memnonic regimes and counter-circuits – which is one reason it had been patrolled by hermeneutic gangs and anti-viral software, for which a courier like Ronell would be a sort of anti-body. It dissolves into a site, a non-site, where diverse temporalities and allo-anthropomorphine traces converge and transit en route to other systems of sense and event. Ronell is the flâneur of an anarchival shift Derrida would tarry before, not wanting to handicap his game – among else, the deliquescence of the era of "the book" as but one dossier and memory regime within the prehistories of teletechnics as such. And, beyond that, the abrupt tempophagy of a post-Anthropocene horizon. She recounts this: "there had been a non-caesaric change. Nobody could scan the cut because we had experienced an interruption in history altogether different from the ones that had been prescribed" (*Crack Wars* 69).

There is no sanctuary from the oozing, which can deluge a delta or coastal mega-cities today – the incursion of the non-anthropomorphic order into the homeland. No interior for retreat, vaginal or green, no *Ursprung* for that matter. No escape. Not if this prefigural ooze is virally embedded as a phantom within every logological sanctum from the concept of "trope" to that of "woman." As the allogendered Emma displays, "woman" as constructed remains a poisoned effect in the archive, a locus of crack war – where "crack" as a prefigural figure opens a "Medusoid rift" across a pan-cultural nexus of caesurae (semioclastic, pharmoco-psychotic, sexed). Ronell: "'A woman' is the mark of a figure in active living, a thing of the sidelines, beside the point and attracted actively to a substitute for active living. (This shows what a symptom woman continues to be, one in touch with vampiric death threats, for what else can a substitute for 'active living' evoke?)" (101). It is delicate for Ronell to expose this technicity of "woman," of literature, to sacrifice it, which one must always perform slightly other than: that "literature" was never other than a pharmacotopic dossier, a power and node, within a teletechnics of which it – or the genocidal memory regimes of "the book" – has been but a signal dossier. In offering her things as "books," engineered objects, marked, embellished, exceeding themselves in typographic shifts, the technology of the "book" is also suspended, and one is delivered over to the non-metaphorics of switchboards, satellites, the "voice" of Ronell (at this point not-a-woman). Exciting "objects," opiates of cognition as well as grenades, they disguise something other – not the immutable stupidity of inscriptions but the fragility of these in a time when old software has run against its limits, drained reserves and degraded biosystems and cognitive regimes. Concealed in the faux folds of this writing, the cracks are telegraphed critico-blogs speaking en famille of the insider's aporia of "transmission" at the point where the success of corporate transit lines fades into runic mockery – a time, let us say, of war, if war always implies a recasting of temporalities.

This project looms beyond mourning, like one trapped too long in the revolving doors of a "transvaluating machine." And it offers itself, or "opfers" itself, in the manner of a faux or painted Isaac who knows better, at the rim of a revoked sacrifice (on the trials of having a psychotic Abrahamic father, see the closing pages of *Stupidity*, cited below): it is almost willing to be erased in the name of something other, or at least think about a suicide that expels the ooze: "Madame Bovary committed total suicide" (*Crack Wars* 94). This is why the trope of the switchboard levitates into satellites – to militarize, it kicks away gravitas and circulates. These also scan the emerging post-global surfaces of a teletechnic Earth whose definitions, and consumed futures, are bound to this system. What occurs when a satellite is reprogrammed, brought down, or worse, understood to have "turned" to the enemy side, become hosted by the black ooze? One is still, has always been here, at war – but with, and in the name, of what other to the red herring called "the other"?

. . .

This rivulet of *petrolepathy* does not stop fusing literary tendrils. One cannot escape *Madame Bovary*, nor the ooze. Faulkner passes *the* book, the book of books (*Madame Bovary*), to another pocket. The trickle of black ooze turns into a proactive agent, a character in the novel *Sanctuary*. Faulkner seems to have sniffed out the implications of Dostoyevsky's or Myshkin's bulging pocket, and panicked, totalized and engorged the encounter. This time, it is passed to a country lawyer with the name of a classical poet, Horace Benbow, a litterateur seeking escape, seeking sanctuary in nature, from his home, from his women – they are out of control and use up or disrespect him, parasites become hosts. He is forced from the house to a "spring" in the country, to nature, where a curious scene of de-origination in American allegory and telemnonics is staged. Here the pocketed volume will precede any return of a narcissistic reflection at the spring, elicits in advance from behind the bushes a prefigural and nonhuman gangster: what is called "Popeye". In a verbal exchange, the black oozing from Emma's mouth is proactively identified, and by smell at that, with the non-figure of Popeye.

As they leave the spring or redneck *Ursprung*, Benbow gives a name to the Poesque bird ("a shadow with speed") that swooped by Popeye, causing the latter to panic and leap, "clawing" at Benbow's pocket, which has been spoken of as having a book in it:

"It's just an owl," Benbow said. "It's nothing but an owl."
Then he said: "They call that Carolina wren a fishingbird.

That's what it is. What I couldn't think of back there," with Popeye crouching against him, clawing at his pocket and hissing through his teeth like a cat. He smells black, Benbow thought; he smells like that black stuff that ran out of Bovary's mouth and down upon her bridal veil when they raised her head.

A moment later, above a black, jagged mass of trees, the house lifted its stark square bulk against the failing sky.

The house was a gutted ruin rising gaunt and stark out of a grove of unpruned cedar trees. It was a landmark, known as the Old Frenchman place, built before the Civil War....
(*Sanctuary* 7-8)[13]

Before what civil war? Our own, today? Black stuff. Popeye, now an animeme, feline, clawing, is linked to the black fluid itself. It reflects the stupidity of American bluntness to name, totalize, to say nothing of smell or personify this techno-mnemonic drip. It is taken out of the pocket, vaginal or book-lined, or identified with where this mock-interior turns inside out, prolapses. The "black stuff that ran out of Bovary's mouth" will, discreetly but inevitably, contaminate every use of the term or figure of blackness in Faulkner (including, especially, "race" – where blacks take on the power of telepathic readers). Thus, the scene shifts to the house full of feebs, the media house of Flaubert, the "Old Frenchman," the shapes and letters of whose name anagrammatically permeates that of "Faulkner." But Emma has, here, in a series of transformations, become the gangster animeme, the cartoon hero-sailor who takes hits of spinach as some drug or steroid. Popeye, on a hit of this technic, has superpowers. The literary has warped, dissolving its anteriorities in a sheer technicity personified – or almost, since, as we hear, Popeye's face is chinless, like wax melted away by a flame, pre-facial. The ruined house of media or "landmark" of the French or "modernist" novel comes from before a civil war, or at least a civil war it is cognizant of inhabiting. If today, this civil war is resurgent, on display and exhibiting weird regurgitations of its predecessor – and if it is all plastered, in Trump's "America," against a restitution of pre-constitutional digital monarchy, a reactionary roll-back of the Anthropocene imaginary itself, that is mapped against an Alamo-stand for a demographically doomed "whiteness" imaginary, a pseudo-geneticism poised to turn against its duped, aggrieved followers first (since the species split being engineered is not that of white Anthropos but the digitally hyper-wealthy separating out, taking whiteness along, for a while). Given this inflammation of the fallen plantation order of white hermeneutics, Faulkner's runaway – who reads and translates "Faulkner" into twenty-first-century writing – draws us toward a resistance within reading-writing of an extended civil war that, in turn, covers over, engages and distracts, harms and acts out, but is staged before a larger question of extinction narratives and mutating away.

Ronell circles back to this zone of the prefigural and tracks it to an impossible, wandering metonymy or non-source. Here that is: Emma, the letter "M," the black ooze, oral and menstrual voiding. But something has happened along the way to "literature," which does not survive itself as an "institution." A question of reading is posed between the lawyer Horace, naming a classic Roman poet, and the mass cultural Popeye at the hyperbolic spring ("Do you read books?"). What Horace has in his pocket, unlike Popeye's gun, turns out to be the "book":

> The drinking man knelt beside the spring. "You've got a pistol in that pocket, I suppose," he said. Across the spring Popeye appeared to contemplate him with two knobs of black rubber. "I'm asking you," Popeye said. "What's that in your pocket?" …. "Don't show me," Popeye said. "Tell me." The other man stopped his hand. "It's a book." "What book?" Popeye said. "Just a book. The kind that people read. Some people do." "Do you read books?" Popeye said. (*Sanctuary* 4)

Do you? Let us suspend addressing whatever a pocket is or may be – or if this hyper-allegory may not be transferred to any reading encounter, any text marking the instant of transmission and splitting. At the opening of *Sanctuary*, the literary seems to recur to a faulted spring or *Quelle* – a Delphic and prosthetic crack. Even if that "spring," here, is as if surrounded by bushes, trees, natural props. Even if literature is being clutched by the lawyer as another refuge or sanctuary, and precisely its female emissary, too, emits a menstrual flow that is eviscerating still – no sanctuary again. The spectrality of literary or even cinematic networks, like animation, is neither a revelation nor a conceit. It is at once a banality and a premise of intervention. It loops back, before arriving, to a prefigural site, which contracts temporal chains and anteriorities. Before the spring appears the law, a country lawyer, Horace Benbow, who seeks refuge from his family, from the storm of controls and betrayals and abjections its logics implies, from his allegorized wife, Belle, who always wants shrimp, to his tarty stepdaughter, who disses him. He flees from "woman" as sanctuary. He is surprised when across a narcissist pond something steps out before any reflection is returned – from behind the supposed natural setting, the bushes. Source of otherness, what greets Horace at the *Quelle* cannot quite be another human. The name and figure of Popeye cites graphic animation, cartoons, as well as a rupture ("pop") of the ocular itself in some sense – here, of reading or memory or perceptual programs. The law that enters here, like the work's own reader, is a refugee from and courier of literary and hermeneutic virals.

Sanctuary, when it was still possible to market itself as "literature," posed as a pot-boiler or prosthetic rape "novel." It hid there. It would not then be read, as the name "Popeye" announces, as a semiotic rape of auratic or mimetological premises (one should say "temples"). At the same time, it or "he" signals an invasive pre-contamination of and by so-called media, a criminalized popular or mass culture ("Popeye the Sailor Man"). It or he remarks a turning out of every interiority or pocket which the piñatas of humanist or Americanist criticism seek to restore – the historial subject as such, the South and its regionalist voice, literary history as a manageable parade of styles. And he does this with a nod to cinematic animation, the technic that graphematically supplants at any hyperbolic spring or *Ursprung* a band of inscriptions inverting the order of "life-death" and with it classical aesthetic assumptions. Popeye's caesura is that of a gangster from Memphis, Egyptoid polis of Tennessee. He or it is the last and crudest American phantom which the frustrated Horace Benbow finds emerging from behind the natural setting – as if from a dominating wife and her shrimp. And like the black menstrual blood leaking from E.B.'s lips, he discloses a world with ceaselessly emptied sanctuaries, corncobbed by this prosthetic liquid coalesced to the black knobs of Popeye's eyes.

Horace Benbow comes into contact with a certain zeroid figure, a black hole named for a cartoon character who slashes the eye of ocularcentric programs, as does all animation, and who comes from behind the bushes – that is, from where a certain faux interior, which is to say sanctuary, was to have been. Faulkner violates temporalities, sheds regional locus as "America" unnames a saturated field – he undoes the hermeneutic race epitomized as a circular and sterile ritual in the pre-civil war positioning, say, in the story called "Was" that opens *Go Down, Moses*. That text or title will name anteriority as such (was), yet reaches into the dead ritual of the childless, white male twins, living like husband and wife, old and faux theophantic binaries, Latin-named and Greek-named, Amodeus *and* Theophilus, "Buck" (fauna) and "Buddy" (flora). The two, sucking into themselves all binaries predicated on abdicating male-male plantation logics, are paralyzed in ritual hunts and returns, before a dawning cataclysm that will rearrange all (civil war). In the short text, the term "race" is also played as a kind of cartoon, slapstick hunt, in which the absurd tracking or reading ritual of "*old* Moses," the hunting dog, chases a pet fox in play around the house, again and again, upending everything. But in a work titled *Go Down, Moses*, where the going under of a certain model of the law or book that precedes its own inscriptions is named, that the stupid hunting dog is named "*old* Moses" gives pause:

> And when they got home just after daylight, this time Uncle Buddy [that is, Amodeus] never even had time to get breakfast started and the fox never even got out of the crate, because the dogs were right there in the room. Old Moses went right into the crate with the fox, so that both of them went right on through the back end of it. That is, the fox went through, because when Uncle Buddy opened the door to come in, old Moses was still wearing most of the crate around his neck until Uncle Buddy kicked it off of him.... and they could hear the fox's claws when he went scrabbling up the lean-pole, onto the roof – a fine race while it lasted, but the tree was too quick. ("Was" 28)[14]

The only occurrence of the name of the ur-patriarch and stuttering lawgiver in the volume so named, named for "his" going down or under, is given to a comic house dog? The hound here is a clownish figure that ritually chases a pet fox inside the house, a practice "race" repeated as ritual. Race, on which the "house divided" is set or faulted, the entire histories of this agon of binarized being (who or what is the man, the human), is referenced to an aesthetic ritual of a loopy hound named *Old* Moses but recalling more Disney's Goofy – all the violence, all the alibis of the word "race" are diverted to a hermeneutic chase doubled back on itself, a dead plantation order of reading.

The war to come here, *the* Civil War, is also the Benjaminian crack wars, for which any so-called war on terror remained a screen and distraction – the faux totalization of the double-chase model as evasion of something else, stupider, lacking aura, more desperate or "material." He or it (Popeye) was here already, and it has something to do with, or at least smells like, the black ooze from Bovary's mouth. A book in Benbow's pocket, mistaken for a gun, could in turn be mistaken for a corncob. Rather than threading discreet, infratextual labyrinths (as in *The Idiot*), Emma appears as gun, then as corncob.

Ronell comments on the implications of Myshkin's carrying *Madame Bovary* in his pocket at the time of his collapse, as of Dostoyevsky's marking a lethal pact between the two performances: "The coupling with the other work seals the suicide pact, ratifies destruction: Emma, Emilia: dial Em for murder" (*Stupidity* 223).[15] Leaving aside the temptation to a diversion to Hitchcock here, we may add, update, Americanize, totalize this call: dial "M" for "Moses," since, as indicated in Faulkner, the pregression of "origins" reaches back before that of the progenitor of the era of the book, and the law, Moses, signatory of an antebellum or plantation hermeneutic. Popeye's link to animation is heightened when he is likened to an electric light, his eyes again black knobs of rubber.

The entire pretext and cache of *interiors* has been corncobbed, together with a network of hermeneutic programs that contrive the return of the escaped slave or animal. We return to the spring. The two wraiths – the lawyer and the outlaw, the reader and the technic other – have a face-off, a non-reading contest even, after which the first will be apprehended and led off by the second. The itinerary of crack wars leads here, all but unthinkably, as if the war machines at the edge of the non-existent "era" from which Ronell writes, with understandable fatigue, know this: one writes backwards of an antebellum era, the plantation hermeneutics of Amodeus and Theophilus, yet in the writing of itself that is already disinscribed – presented as a sterile plantation ritual, as Faulkner writes it. Popeye will be in control of everything as a Memphis or hieroglyphic gangster allied to teletechnics and animation whose frenzied impotence leaves him whinnying vicariously over Temple and Red's performance in a brothel. Temples, nature, spring, enclosures and pockets of all kinds, "sanctuaries" constructed over an occlusion – all in advance violated and dis-interiorized. The book in the lawyer's pocket guarantees this. Horace does not want to name Madame Bovary, the woman he takes with him to escape from his women; not knowing that the inverse model to himself that Emma incarnates lethally voids the literary "hit" he seeks. The personified "ruin" of a mediatric house is permeated by latent horrors and cinematized crime, impending murder, stupid folk ("feebs"), while conjuring Faulknerian or American writing tout court in its faux modernist moment. Old Frenchman's place is an eviscerated structure housing impending violations that epitomize what remains for Popeye, amid the crack wars of prosthetic romance. It defers American or faux modernist writing to an anterior colonization – or simply an anterior, allolinguistic contamination.

What Popeye incarnates does not itself read. He is illiterate, like the shrivelled telepath, black Aunt Mollie of the tale "Go Down, Moses," who is content to stare at a newspaper article about her dead nephew though unable to read. One is in the "American" weave and trance placed over Natasya, a metonymic corpse needing disinfectant to stop the viral takeovers, over the suicidal pact Ronell inspects between Fyodor and Gustave. With Popeye's appearance at the spring, all of "Faulkner" the regionalist, the southern "writer," the historial "American" voice, and so on, finds itself as translated back into the black flow of a bile linked with this "first" modern novel to sheer anteriority.[16] All referential rites are as if liquefied here – including, under the shadow of Old Frenchman's place, the "American" as such. The detail of Bovary's corpse suggests a literature that begins "after" its institutional death and faux mourning: and with that, the archive is exposed like New Orleans's mausoleums' coffins after flooding, opened to re-inscription. Naturalism, historiality, psychology, character, ocularcentrism, symbolism – whatever you like that is, still, auratic and blind – is sucked into the (blind) reading of Popeye's black knobs. The zero covers a fault where the system has as though been corncobbed itself in advance, in a perpetual trance or aestheticized narcosis.

What *disinfectant* of what viral discharge is or is not covered, is covered and uncovered by this "American clothe" or weave?[17]

Here Ronell stages a different sort of outing – the test drive, (s) he calls it. In tracking the prefigural ooze from Emma's mouth, stupid, amaterial, inscriptive, a detour is activated. In the "Rhetoric of Testing," a cipher chapter for *Stupidity*, a family plot of dead-enders is visited – de Man, Benjamin and Friedrich Schlegel:

> The welcoming of irony and allegory, as Schlegel's text indicates, is the kiss of death. For not only is there an impertinent emphasis placed on the non-convergence of any stated meaning and its understanding, but this engagement lets loose a cannonade of demystifications that can ruin a career (the poisoned Socrates, abjected Schlegel, flunked out Benjamin, dead de Man, et al.) or, at the very least, exacts revenge in the form of a total religious conversion. (*Stupidity* 159)

Begun as a precarious genealogy of stupidity, of cultural transmission and resistance, these figurines end in a kind of swamp orgy, a frog-froth of sterile power. With Popeye, Ronell implies this is being played out in the anteroom of a step beyond, a "test drive" mutation to which all these legacies point or catalyze. Thus, for Ronell, one would be perhaps already outside of this – those who touched the live wires of "allegory" and "irony." She asks after prophetic mutations in or from this legacy, unmapped speech acts: "To what extent is the prophetic word indebted to irony? Can there be prophecy without irony, I mean in a nonpsychotic sense?" (157). There is a short transit between the vacating of "irony" as a specular infinity to something else, called "prophetic," piercing future timeloops and collapsing temporal columns. Ronell knows that knowing knows something against itself, at war with itself, that the political, today, involves only epistemological horizons where alternative programs of memory, sensation, reference, consumption may be set – that these are anestheticized, perhaps paralyzed by everything the drug hit (of all sorts) would oppose or counter-rupture. Ronell implies this is being played out before a step beyond to which all these legacies testify: wars of re-inscription. Crack wars. Wars always lost.

How does Ronell, less and more than a tele-flâneur, more or less permanently fatigued, prepare for such wars? How are they, still, anarchival, proferred as if at the non-site Derrida names *khora*, betokened by an efflux of black bile, where script has been returned to ink? How does (s)he repeal or evade the anaesthetizing traps? By stepping into them and pretending to be vanquished, then finding the pool of stupidity where the police don't want to patrol, the spring – then setting up para-networks of communication for the readers-to-come who may need such in place? Primitive trope in its way, the satellite nonetheless revokes metaphor. Kant is such a "satellite." So too could be "Plato," "Faulkner," and so on. Satellites protect the stratosphere of informatrices and faux perceptibility and mnemonics. They manage signals, or histories. And they can crash (Houston, we have a problem...).

Ronell has wandered into a charmed spot or non-site of *ex-scriptions*, rehearsing rituals of criticism for lack of a genre. All of her weapons are fronts – literary history, the frog pond of the politics of transmission, self-dramatizations without self, the *stupid* insight that cannot be given shape or name but is everywhere enshadowing. (S)he has stepped out of the room, the house, and is left acting like she wants in or back (a motif of expulsion runs through her text). (S)he paints the edges of her pages with sparklers and rockets, tropes philosophical names as "satellites," embraces stupid voices she rescinds ("De Man"), clowns too much, probes imbecile victimages, gossips inside stories to the dead – as if the figure of the clown the *Genealogy of Morals* recommends as strategy were a tarrying site. (S)he sustains and rescinds these histories which she experiences as accelerating circuitry, buzzings that return to and evade this ooze. One expects, among her catalogue of de-extincted pharmacopes and illnesses, her cornucopia of anarchival poisons (a *grand mal d'archive*) to encounter a certain palsy – as where an arm or shoulder, part of the mnemonic body, enters a dead zone neither subject to shaking, nor trance, nor mock-jouissance, nor life-death. (Ronell, in this sense, parodies the aura of the female poisoner.) The unleashed digital dogs of coming wars of reinscription are like "old Moses." Everything in Ronell that testifies to and reverse-accelerates against to maintain a position of articulation, to not be targeted, to maintain the freedom to hang up the phone, to ironize the non-call that is implied – these ooze between the lines and direct Ronell's syntax. Ronell offers cognitive blogs even while "literature" is revoked. And one wonders, in this machinery of indigestibles, what

Odradekian monarch would emerge, like Myshkin, in sacrificial ardor or self-extinction, if these strategies run out: what would speak, be instantly abjected, unrecognized, then return otherwise when the tropology subsides of the necessary chatter of short-lived aesthetic histories like classicism, modernism, humanism – when they suspend their pretense even to be ghosts? She, unmarked, reads from the era of oil and biomorphic going under.

This question underlies the meditation on Abraham and Isaac that closes *Stupidity* – faux son to faux father, at the limits of familial travesty. It exemplifies the "Medusoid rift" Ronell can situate herself-himself in, probing perimeters of totalized political spells in the archival orders. Apotropaic, this rift refuses orientation, turns toward the dumb, the mute, toward the mutation of inscriptions that set the levers of perceptual and temporal regimes. If Derrida seemed to morph from the analytic of hospitality to an ill or fevered archive and then to a "suicidal auto-immunity process" that makes the house itself a self-canceling structure, Ronell drifts outside of the metaphorics of the house, or the family – and looks back, puts on the costumes to see. *Stupidity* speculates on "this that expelled you from your house" – or its pocket – in its final line, after dramatizing the dilemmas of an Isaac "cheated by the call":

> Assuming that Abraham was cured and did not sacrifice Isaac (though according to one midrash the son was executed), the question remains of how and whether Isaac survived the near-death experience – how he survived a psychotic father, that is, everybody's primal father, Kafka's, yours, and mine, even, or especially, when they are in sync with the Law. (309)

The "psychotic" father is the one without paternal identity (which is never biological), who disowns his own premise and may be, in effect, a woman: (s)he undoes the pretext of the familial – that is his open secret (like Derrida's "I am not of the family" amidst the projection of the opposite). But the reference to the one exception among what may be called the "Isaac Variations" is telling. It is to a midrash that says, "Yeah, of course, he was sacrificed and all the speculation covers that up for a reason": if Isaac were cut off, as he, of course, was, the future would have never taken place, and the "present" occupied by the commentators revoked as spectral. The entire memory system by which a certain model of the house or "present" is maintained would turn out, in essence, to have been an implant more or less self-canceling, accelerated by having been programmed on "real." Eyeless in Gaza. And, of course, a kaballist reading – the midrash Ronell mentions – asserts one cannot assume Abraham was cured, that Isaac was sacrificed, that the generations of which one is a part since simply did not occur – and that one persists, today and already, under the cloak of the already extincted but yet to catch up, of participating in a mode of de-extinction and cin-animation we mistake for "life."

The black stuff or ooze, ur-material, stupid, inkish, khoratic, a prefigural agent with prehistorial properties – it exceeds, it precedes, it contaminates. It leaks through cracks. Perhaps this trickle, which enters Ronell's calculus and possesses it, is about a catastrophe for which one needs to develop a new vocabulary – one that is not monumentalized as past trauma but oozes, proactively mutates and consumes histories, beyond mourning. Ronell's writing can simulate a Shoah's ark redux, the viral gossip of a theoretical afterlife pretending it does not know, at every turn, the extinction of the genres it mimes itself out of. No wonder "Ronell" is fatigued – even where she uses that as a front or day job, like Popeye before eating his anything-but-organic spinach. But then, once the referential spell is cracked, reading comes to this site which is irreversible.

Does the "Alt" version of Isaac as indeed dispatched, cut off like a sheep offering, which is to say taking with him all future generations, suggest an alternative reading apparatus – over the occlusion of which "we" subsequently proliferated *as if* anyway until circling back.

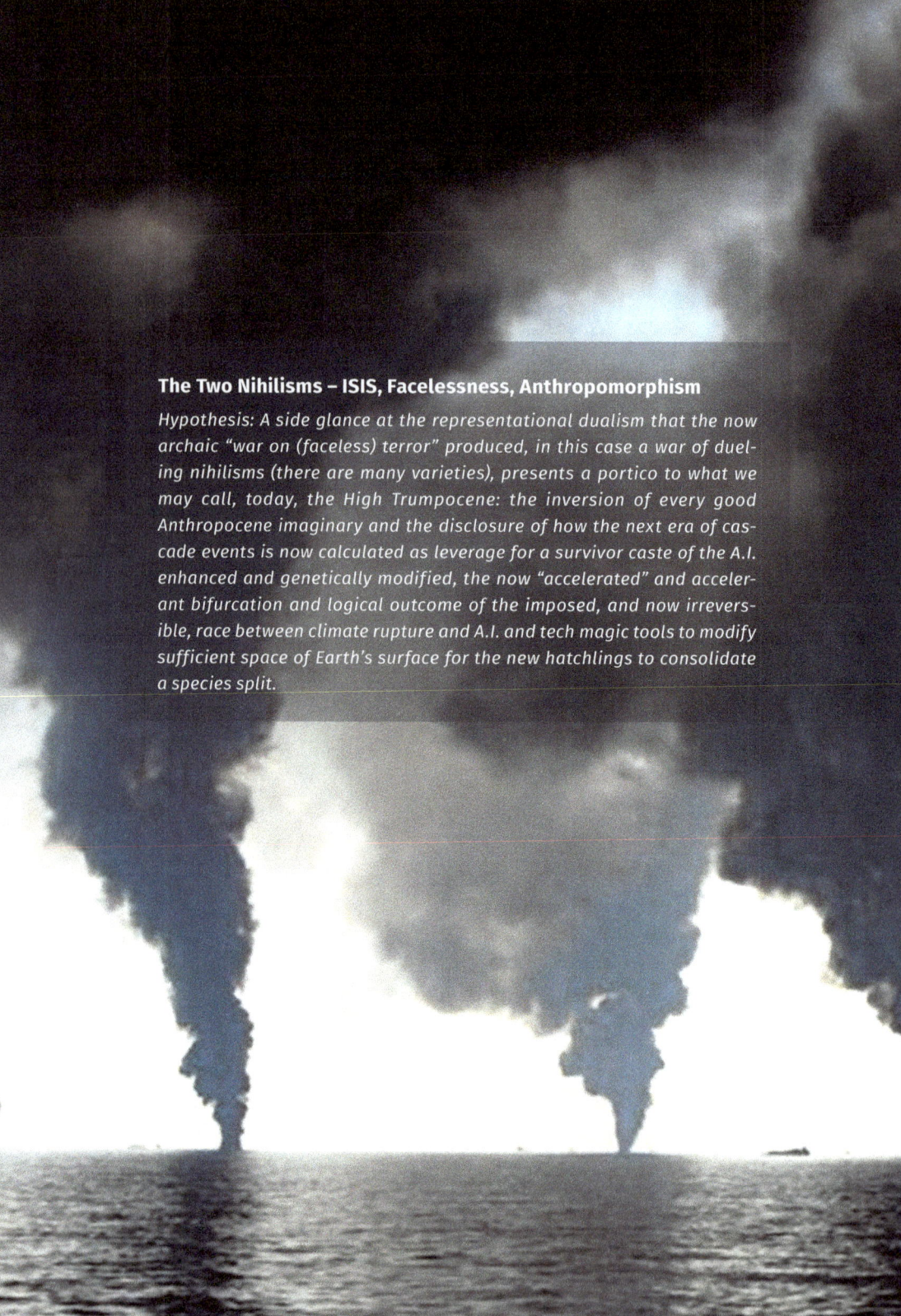

The Two Nihilisms – ISIS, Facelessness, Anthropomorphism

Hypothesis: A side glance at the representational dualism that the now archaic "war on (faceless) terror" produced, in this case a war of dueling nihilisms (there are many varieties), presents a portico to what we may call, today, the High Trumpocene: the inversion of every good Anthropocene imaginary and the disclosure of how the next era of cascade events is now calculated as leverage for a survivor caste of the A.I. enhanced and genetically modified, the now "accelerated" and accelerant bifurcation and logical outcome of the imposed, and now irreversible, race between climate rupture and A.I. and tech magic tools to modify sufficient space of Earth's surface for the new hatchlings to consolidate a species split.

3 Thirteen Ways of Looking (Back) at ISIS – or, "Jihadis 'R Us"

> ISIS; or what is void mastery and blank control? What is orange insurgence? Chromatic hell? ISIS is – ISIS – what does "ISIS" mean? What kind of sign is it? We treat the world like we think ISIS treats us like we treat animals. Can that be serious? Can they be? In good faith? And we?
> …. Are ISIS something like an ontological breach, a sort of liquid time machine of actual pain way beyond what we think – if we want to attend to 'the victim', then let us really begin with the most difficult and spectral one…
> —Jonty Tiplady

In 2025, all the recent ghosts and invaders seem to return, if not in cartoon forms and deep fakes, then as a kind of swirl of tele-mnemonic debris or tooled regression. As ISIS has returned as a now background agent and enemy worthy of attack, it may repay looking at how they had figured among the black-hole imaginaries of the pre-Covid times, returning today as background player, both woven into and a distraction from climate chaos's dawning irreversibility.

There is reason to inspect the media traces of an enemy phenomenon that altered the geopolitical tilt (the eponymous "war *on terror*" that shadowed the unfolding post-9/11 debacles (Iraq), but do so in a somewhat melting Dali mirror, and against the unfolding of climate chaos and the *panic of reference* it induces globally – hence the herding into the sprayed hermeneutic popcorn and pretzels of generated conspiracy micro-narratives, instantly replaceable, or the collective occlusion of the overriding maelstrom of biospheric mutation and mass extinctions. With this in mind, one must put on night goggles to track, as an aesthetic doubling between what we may call the two hyper-nihilisms. (There are others, and we await a monograph separating out and hierarchizing these.) As such, "Isis" – apparently on the rebound in 2024 – presents a mini reading puzzle of the *Nethercene*.

Among the image debris of the mid-2010s were the twin media antibodies of ISIS and Ebola, chosen to channel a lethal outside leaking through the telepolis's shifting defenses, mirages, cell walls, borders, perceptual totalizations. (Let us mark this "date," by the way, as when tipping points have officially passed, and something like a vortex is entered.) Read as what Jonty Tiplady above names for us "an ontological breach," it dissolves borders, the incarnation of a sort of black ooze or "sort of liquid time machine of actual pain." Perhaps one can only wander within these sentences, as in the limbo the church assigned for dead infants to percolate in in perpetuity. Like Ebola, an invisible microbe that liquefies internal body organs, ISIS would invasively take over cell or town structures as a black, hooded pool of assaulting memes. Or so the metaphor, and media effort, to interface two things without any overt connection whatsoever – except as they both relate, perhaps, to climate change. Both flow as from some other side glimpsed at, recoiled before, by the collective personhood of the West's "we." With so much pop cultural wit to sidestep, to say nothing of geopolitical and sectarian web-weaving, I'll focus only on barely two of thirteen (hundred) ways of looking at ISIS. These are double, and point to the blackbird effect of a poetic impasse: on the one hand, it appears as a defacement of the "Anthropocene" and the Western epistelarium, one that, today, accelerates ecocide behind post-Enlightenment façades; on the other hand, it is an effect called forth by the latter.

As far as this goes – and one would need cipher in the return of ISIS today – ignore for a moment the proliferating climate disasters swarming continents; and ignore the spell of the telepolis itself, the cultures of "distraction" or the re-engineering of a klepto-feudal oligarchy more or less corporatized and without place. The "ISIS beheadings," expertly made videos both admired and analyzed as steeped in cinematic culture, or tricks, and angled at video gamers with the lure of transposing screen fantasies of medieval slay-fests into the "real," merit a pause. Er, beheadings on YouTube? These heads – one of their own, even carved by one of their own (British rappers, and so on), spawned a new war, coerced by public response to the insult of insults, having your own reporters decapitated, heads appearing on chests, as if looking back. It is not that the thing-nature of ISIS used the West's mimeticism against itself – where hundreds of thousands of torn, gassed, mutilated and destroyed Arabs do not rouse empathy, but one face "like us" in the camera. The West's decapitation as if by itself echoes that its achievement, the Anthropocene, inescapably involves mass extinctions and ecocide.

It may be an error to regard ISIS as a radically alien force. In a way, it represents a technological evolution of, and bifurcation against, globalist and Western DNA, and an inversion of a certain artefaction and totalization of that "life." It appears as a growing black stain or "ontological tear." Just as it leaps from alienated video gamers zapping medieval screen creatures or combatants to crucifying and beheading real bodies, ISIS defaces explicitly the mimetic spell that defines the Western cartel's extractivism and ecocidal accelerations. "ISIS" as sign – as a compilation of media effects – seems momentarily readable as a super-organism and a defacing counter-modernity.

Thus, in the sort of hyposcript of the moment, well below any "climate change unconscious" which traverses all discourses today, ISIS melds with the nano-terror, inhuman, viral, invisible, traversing all cellular borders, without an "other" that the critical imaginaries of last man culture can masticate on and try to place or grant personhood to:

> The murderers are often young men from the West, so enraged at life that others end up losing their lives to them. ISIS is something like a reactionary globalization vanguard whose goal it is to subjugate the planet and welcome all miscreants into their fold. Ebola too – the other danger that could pose a direct threat to us – is masterfully ignoring all containment efforts.... A virus like Ebola doesn't care what part of a city it's in, and the jihadist doesn't have to invade – he's already there. He's a former boxer, a rapper, a good student, whatever, and then he becomes indoctrinated by some stuff he sees on the Internet.[18]

ISIS, however, has to compete with Hollywood ultra-violence – its aesthetically admired video editing, debated for its cinematic citations, turns the screen against itself. It does this, of course, by literalizing the former in blood and the recorded defacement of inscriptions. The black-faced figures would stand outside of "light" or individuation or any contract with face. That is, outside or before figuration itself. The plethora of military, tribalist, historical, Islamist, "terrorist" and culturalist analyses or mock-ups testify that it elbowed its way into the pop-cultural A-list by bleeding into an aesthetic event – that is, one that protrudes as the defacement of any "Anthropocene" it out-nihilates by literalizing the screen or positing inscriptions without any representational corollary. There is no "other" that a Judith Butler can commiserate with. In fact, the fiction of a single "humanity" is defaced, another "Enlightenment" placard fronting for klepto-oligarchic cartels and techno-psychic hollowing.

One cannot *read* ISIS outside of the loss of face or aura of "the West" in the catastrophic Bush wars and the 2008 financial "crisis" – or the import of irreversible ecocide deriving from its hyper-industrial monoculture. ISIS plays head games. It gives rise to dissertatings on "beheading," as ISIS's YouTube shockers literalize a pop-cultural imaginary we already turn to reflexively, and absorb on tele-circuits (*Game of Thrones*'s heads on pikes). ISIS promises recruits can "live" the medieval fantasy wars disaffected youth video gamers practice daily. ISIS defaces – not only the ancient temples and cultural treasures but the tele-streaming totalization that posits itself as hyper-industrial modernity, "the West," yet whose teleological fantasies project escapist singularities arriving. In the battle of the two nihilisms, ISIS is faceless yet gives rise to a chain of citational cowbells. Žižek is disappointed in the fake "fundamentalism," seeing it as envious of what it rejects – even if that is the comfort and security and sexual ease of last man culture:

It may appear that the split between the permissive "First World" and the fundamentalist reaction to it runs more and more along the lines of the opposition between leading a long satisfying life full of material and cultural wealth and dedicating one's life to some transcendent cause. Is this antagonism not the one between what Nietzsche called "passive" and "active" nihilism? We in the West are the Nietzschean Last Men, immersed in stupid daily pleasures, while the Muslim radicals are ready to risk everything, engaged in the struggle up to their self-destruction.[19]

Which returns us to a "reality TV" snuff video itself – the innovation that got ISIS A-list pop-cultural status (no doubt, it can label products to raise cash when it wants), admiring analyses from cineastes, and forced Obama to do the last thing he wanted to, open yet another Middle East war, particularly one the liquid and techno-evolutionary ISIS could, like Russia's "hybrid" war, simply win. The difference with ISIS's snuff videos is embarrassingly banal: they require the identity of a known individual's name and known face, in journalist James Foley's inaugural case that of a journalist, a reporter, a media-affiliated proper name. There would be coming attractions to maintain suspense and seriality, and something of overkill when, this becoming familiar (oh, another beheading?), they turned to fire. The crude selection of hand knives over swords for the butchery, sawing rather than a clear cut, further diminishes the status of the killed to that of farm animals. Here, on the video, there are no "humans" but ISIS – and they are not "human" in the Western sense, are without face or personhood, a reverse anthropomorphism.

■ ■ ■

The West relies on face identification, an economy of face out of which their cognitive, legal, and referential regimes circulate – and specifically one of their own. Biometrics and surveillance matrices identify movement and map variants. So this arrives on YouTube. Hundreds of thousands of Syrians barrel-bombed, massacred or tortured didn't register much or else settled into infotainment from the televisual peripheries or doomed zones. Sawed off by one of your own – British ex-rapper, "Jihadi John" – the West's head is cut off by itself, by hooded Anglo offspring, the *capo* itself then as if looking back – after reciting to the camera poison scripts against itself. ISIS's video, though, recalls less the editing techniques of *Natural Born Killers* than a sort of Islamist *South Park* – and it has been, accordingly, marked by being montaged with the latter's hooded pre-schooler Kenny, who routinely is killed and brought back in the pop-cultural perpetuum of the latter's child limbo. The black anti-matter of ISIS, liquid and adaptable, stands outside the globalist infrastructure while feeding from it directly (its oil, its internet, its acquired weapons, its funders). This gives the media effect called "ISIS" the status of rejecting aura and personification, as if on behalf of a total outside to an ecocidal trajectory that feeds off disposable populations and territories.[20]

ISIS excretes something mock-romantic. It liquefies and defaces the mimetic regime, piles corpses decimating any cultural identity, positing a "new" world voided of the contaminants of the West's fragile and enslaved meaning industries. Prefigural, ISIS manages to simulate a leakage of black fluid, an anthropomorphism of oil, responding at the site at which climate impasses would be arriving across the "Middle East" (Syria's collapsed agriculture, Egypt's bread riots). This liquid, inhuman and inanimate yet violently erasing trace practices a visceral defacement of the "Anthropocene" as a now iconic trope.

But ISIS slips the reins – in destroying an economy of face without supplanting that, it displays that "economy" as defaced, void, destructive, indifferent, self-fetishized, dead, worthy of death. This is its appeal to "alienated youths" (web-centric). Yet it also exposes that face was never quite there, was always artificed from a position of originary facelessness – and that, too, is already present in the corporate totalitarian model. It is not that it is a mirror to "the West," injecting their snuff-film into media currency, offering that as a serial.... The blackness of the ISIS hood, accordingly, is not the same blackness it seems to signify – that scary one in the absence of light. Rather, it precedes the binary of light and black as the former's absence. It simulates a technic in advance of that managed "fiat lux," that which produces "light" to begin with, non-anthropomorphic. One is stuck with this conundrum, a parade of freak murderers who flash up as the defaced anthropomorphism of, say, the Anthropocene as an era of extinction and biomorphic collapse. As a cult of ultra-violence, ISIS appears to shred any recognition of the human person at all – it is anti-human, anti-every other sect or creed or cultural identity or "we." It is without a "we," a black stain of memes. In this, too, it mimes the West's exclusionary premise and goes hyper with it: they mime the criminal violence of an *anthropos* that could never equal the "sign" of its Enlightenment premises, any more than Kurz can, since its constitutive "we" implies not universal humanity but invariably the opposite, the imposition and maintenance of an exclusionist definition. Defacing the "Enlightenment" memes together with "the West" from within the former's own toolbox, disregarding it as mock-worthy or contemptible, one can see ISIS as a reverse-anthropomorphism. That is, it appears thinglike in human form, black-attired, faceless, Borg-like. Viewed as a zombie apocalypse by the West, already defending itself from the climate refugees its feeding system created, as anti-"life," they view the reverse – something ill, if not deadly, and utterly self-incapacitating, in the hyper-industrial West, already ruled by personified corporate entities, nonexistent, nonhuman. The passion of sacrificing "life" before its beyond involves a weird echo of Western "meaning" compulsions – only the inscription would not be, "Here is (was) the 'Anthropocene,'" but a demolition of all inscriptions (museums, ancient sites, world heritage monuments). "IS-IS," iterates Tiplady, as if a stutter of conjugation were mocked by the iteration of "existence."

ISIS in this sense generates a substitutive chain of meanings, the thirteen hundred ways of looking at ISIS the media treats us to – all licit: that includes ISIS as a mirror of the West, ISIS as fake fundamentalism (or "Islam"), ISIS as a cult of ultra-violence, ISIS as a global viral dump of hyper-modernities (by way of that). And yet, one returns to question what kind of "sign" ISIS is or works, linked to a prefigural zone of the era of "climate chaos"? The gesture of defacement in that of the beheading – which also throws back and trades against the mimeticism by which the Western stupor and coded violence is maintained – makes ISIS less an anti-Anthropocene eruption than a pressure valve for the latter's consolidating regime (to contain is not to destroy). But there is another weirdness here, which is that it seems the early twenty-first century is going through a mnemonic spin cycle, as if it collectively hit a road bump so void and dismembering that it would turn to pop-cultural tapestries of mythic pasts to clothe itself – time-traveling. Putin now stands for thirteenth-century orthodoxies, Xi for "cultural revolution" light, a more or less zombied America can't remember what medieval feudalism was and tosses out "gilded age" to cover corporate totalitarianization – and ISIS, returning to the rampaging hordes of yore, who also form the safety valve for pop culture and video-game worlds (with its medieval fantasy violence and practice of killing as a disappearance from the gameboard). Thus, the "caliphate" as a deterritorialization not of states but of "territory" and borders of any sort – liquid, viscous, absorptive, viral, Ebola-like.

This reflex to cover the "present" by artificing a mnemonic chatroom and private media ecology (this "Russia"), is mimed, however, in the death-throw efforts of critical theorists, left intellectuals and environmental activists broadly as well: one goes back to "affect" or humanity, to phenomenology or data, to historical memes and political faiths, essentially treading water.

ISIS managed to create itself as prefigural agency – and, quite possibly, that will have been the peak of its achievement and artifice, regardless of the destroyed armies and communities to come (it cannot but lapse toward being a "state" having undone the modus). Is it more "nihilistic" than the invisible acceleration of an economics of extinction by mediacratic cartels and hyper-elite?

I Jealous Neuro-Divergent Nihilisms

This may be the gift ISIS provides the West – a polar vortex of nihilisms. ISIS's über-nihilism cancels the privileges of hyper-industrial and ecocidal nihilism. It is at this point that the loincloth of "Enlightenment" templates are yanked back into place or reasserted against the terrible non-futures implied by this project (Žižek). But one point ISIS manifests – as does "Putinlandia" or invariably China – is that the time-space window of the Enlightenment codex and post-war "international law" is closed, having already delivered the financial crisis by gaming its own "secret sauce" system and the earth to accelerating ecocide. From the point of view of the playing out of species-life over aeons to come, and the future course of devolution of life diversity on earth beyond that, a mere thousand-year reign of arrested medievalism might be just the thing for a remix and ecocidal pause: at present, the approach is to eke out some more decades, here and there, to extend as long as possible the present gameboard. But since we are today speaking from after tipping points have passed, there is a key opacity here. When it comes to the Western critic's favored hobby, that of projecting and incorporating "others" and "othernesses," this appears an otherness to reject, sending us reeling back to the "Enlightenment" buffet we are supposed to reconfirm as a legacy represented, today, by financial collapse and engineered mega-debt, accelerating extinction events, and the deferral and occlusion of addressing ecocide directly. As the faceless black figures of ISIS arrive as a reverse anthropomorphism, putting the latter, and thus the Anthropocene itself, into question.

Now, one must pause – since all of this is also theater, spectacle, including the wreckage of identification and affects. That its military aggression was networked by former Baathist officers of Saddam, intelligence specialists included, hardly fundamentalists, raises the prospect of an opportunistic appropriation of jihad along the lines of Putin's Ukraine adventures, a mimetolical war.

Any purported "Anthropocene" would depend on anthropomorphism, not because it is the key organizing trope out of which human consciousness assembles linguistically and holds the world it conjures in grappling hooks, but because it may not be a trope at all. So suggests Paul de Man, who observes it cannot be called a trope, even though it appears structured like one. Instead of participating in a substitive chain, like metaphor or metonymy, it "freezes" any tropological system: it precedes face – which, the signature of trope, is frozen. Instead, it represents a "proper name," a violence that does not require identification or relation, one structured necessarily on exclusion: "But 'anthropomorphism' is not just a trope but an identification on the level of substance. It takes one entity for another and thus implies the constitution of specific entities prior to their confusion, the taking of something for something else that can then be assumed to be given. It is no longer a proposition but a proper name…." ("Anthropomorphism and Trope in the Lyric").[21] At the core of any "Anthropocene" hums an anthropomorphist software, the sort perhaps now managed by digital corporatisms, Disney, Hollywood, telemarketing and security bots. Since it is not a trope but a "proper name," its imposition is totalizing: it freezes (or totalizes) the order of figuration and needs to exclude, first, all others, other animemes, hominids, tribes, groups, clans, features. The anthropomorphism that upholds ecocidal "Enlightenment" templates and projects a single "humanity" makes any general humanity precisely impossible – it can only impose itself at the expense of its "universality." Thus, the only seeming contradiction of a planetary population that, unable to respond to a joint threat (because it is itself), fragments into competing tribes, enclaves, identities, memory systems, and so on. Thus, the dilemma of the sedate West, the enclave of last man culture, where the imaginary of a hobbit shire is crossbred with and depends upon sweeping violence and mass death in territories or peoples on the telepolitical peripheries – and any species it can consume.

Of some thirteen ways of looking at ISIS, like a blackbird that provokes different relational attempts at metaphoric appropriation, that of reading it as a sign (what kind of sign?) stands out. Which does not mean what it is a sign of – but reading it as exposing sign as noncoincident, first, with whatever it generates as "meaning," the expansion of inscriptive logics to the point of eradicating real inscriptions and all "human" "we's" or cultural formations outside its facelessness. Aesthetically, it has simulated the non-site of what the Western imaginary circles or names as a destroying sublime – further confusing it. This is not the ISIS of radical Islamism, not the "fake" fundamentalist (Žižek), nor the disfiguring mirror of Western behavior, nor the alternative-modernity warriors, nor the video-gamed, disaffected global youth, nor that analyzing an aesthetics of beheading, nor even its positioning before the "Enlightenment," hyper-industrial betrayal of "life" and disposable castes and territories. Rather, any of these senses responds to the trick that the beheading videos executed – that of entrapping the West in the mimetic regimes that sustain the Anthropocene mirage. Our resilient liberal Left, looking for some way to belatedly address or appropriate "climate change" (which really spoils utopist imaginaries), has decided that focusing on its first human victims would be a good way to maintain their purchase on empathy and supposed emancipation. ISIS as "sign" entails the sense of a sign itself, its structure of never equal to itself or organically related to whatever it produces. ISIS signifies what stands apart from a disappearing economy of face, or anthropomorphism. ISIS produces a faceless "other" that cannot be identified with, given voice or face, appropriated, or empathized with – and that would annihilate bodies bearing other proper names (Christians, Kurds, Shia, other Sunni). It is not that an empire of face, tied to economics and a corporate economics of extinction, gives rise at its moment of totalization to a vast defacement of the entire façade, but rather that the norm of facelessness is sufficiently revolted by the parenthesis of face culture that arose as a hypothesis – that of the Greek *anthropos* – become ludicrous and costly (mass extinction events). Decimated, ISIS recurs, as again in 2024.

II Head Games

At the current juncture, though, ISIS demonstrated its prowess in purveying a new form of nomad corporatism – one whose impersonality inverts that distantly of the West in turn. In the end, the face and head of the West it thinks it can game and decapitate were already digitalized or acephalic, more spectral than black balaclavas. "The West" lost its head long ago, and cannot even locate where "capitalism" is any more? And so the appropriate rejoinder to the "barbaric" horde of ISIS and its nostalgia in the gaming world for palpably bloodier times is not that to which the philosophers Derrida and Žižek return, like Indian scouts who make it back to the fort, to the Western ideals to be defended against a rampant death drive. It is to remind ISIS that it can take cities and massacre scores; it can murder and enslave sects; it can practice creative destruction in the desert and make YouTube beheading videos as sport – yet when it goes up against the West, it is going up against a destroyer of worlds and mass extinctor of life forms. And one must recall, ISIS and Ebola were the faceless celebrity threats, distracting from notice that tipping points had passed. Thus, the feeling of boredom behind all the memes and insight cast up, and the sense that a Sony contract, HBO script, or label run would loom for any other such breakaway talent that can techno-evolve and engineer faceless celebrity on iMovie: this return of Saddam's rudely decapitated ghost for revenge triggers, unexpectedly, the emergence of a new semio-species without face, a new corp(s)oratism, free of Hobby Lobby culture's pretense of corporate personhood. Updating this ontological skirmish to 2025 witnesses the spectacle and stupor inducing speed with which this corporate personhood – fueled by A.I. and mini-techno brolets swarming, recoding and downloading the US Treasury and security "secretes" gratis Musk's predatory descent as Trump's enforcer (or vice versa), one is left with a bit of an toss-up in the online betting parlors: may the better hyper-nihilism win (or not). That which would eviscerate modernity or that which would accelerate through all tipping points to leverage climate chaos against the retirement of current species iterations (genetically sloppy, unneeded on Musk's Mars).

But the poisoned gift of ISIS – since all of these attacks, Žižek murmurs, are perhaps a sign of love – is releasing the West from its fantasy of a single humanity and its resulting monocultural trance, or embedding it terminally in such. There is no way the said West accords human status to mass killers, mass rapists, who crucify children and saw off heads, and so on. And yet ISIS, who ostensibly recognize no other group or identity, including within Islam, as sufficiently "human" or sacral, and butcher them, troll the structure of anthropomorphism, which can only appear in exclusionist logics. The good news is we are free to track and disclose innumerable variants that amount to para-versions and subspecies of humans today, not predicated on "race" or geography but on a spectrum of cognitive, neural, referential and mnemo-technic regimes. That is because the artefaction of a "human" is neither descriptive nor refers to a species nor a hominid nor an organic body – we have no trouble assigning personhood to corporations. It is in each case a linguistic invention.

And yet, ISIS spread like a stain. It simulates a war of inscription before any phenomenal or perceptual regime is in place with an order of digital tropes, totalized, and placed on maintenance or autopilot – but, again, behind the distracted and obese consciousness-factories of "the West" lies a more sophisticated and globalized version of the same techniques, the seizure and manipulations of inscriptions to generate the palliative screens of the West, their talking heads. And as far as borderless and stateless corporate entities go that infiltrate and appropriate populations and power formations, or, for that matter, field mercenary armies (paid in petro-dollars or dopamine), the multinationals are laughably far ahead. The culture of last men that is now angling for protection against medieval hordes and anthropomorphized oil even as it rotates back into a klepto-mediatric and neo-feudal caste system – this, just as the world accelerates into serial climate catastrophes and megadroughts – may find itself in a peculiar position as it reaches for more Dorritos. Rather than being the population struggling to preserve the spirit of the West and Enlightenment memes, rather than being the "humanity" that is being protected to keep open a progressive future, they may be relegated to watching on screen the protracted superbowl of the two nihilisms and the two corporatisms to come, both of which regard the other as disposable. After this, everything may be read through the politics of managed extinction, or extinctions, to come. Welcome to the Nethercene, which never arrived yet is patrolled by drones, opening innumerable funnels of time shoots, the time of the financially engineered species split the twenty-second century will talk about – the heirs of the ".0001 %" following the water wars and territorial culls.

■ ■ ■

But lest we swallow the hype, there is some question of ISIS's spectral influence, or whether a certain cultural facelessness in the West does not precede its balaclava minstral show. In an essay in the *New York Times*, Stephen Marché analyzes a new "epidemic of facelessness" in what we will call our "today" – in this case, it enters through a quirk on the periphery of our tele-sphere with a virulent logic. He is analyzing the effects of trolling on the internet, targeting web prey anonymously and with the unrestrained vengeful or defacing drive to violate. The link is not made to roiling para-movements on the seeming geo-political peripheries, like ISIS, with its Ebola-like spread and complete defacement of any residual aura or credibility of "the West" or hyper-modernities. This faceless attack mode penetrates the real despite its ambiguous status – including law courts and jailings for discovered miscreants. "Trolling" operates in the structural logics of terrorism. It can always be dissociated from by a legal individual who adapts a voice or persona and be called all "in play," as if in quotation marks, a form of gaming virtually solicited by the social media networks. The dependency of such networks on photographs, or "selfies," only conflates the dissociation, as persons obsess over representations of the face (unable to reassure enough or leave evidence of having been at a specific space-time). To situate today's "epidemic of facelessness"

– the trope of illness pleads, at first, for a treatment – Marché presents a narrative in which face has been the foundation of an epoch of law, identity, and above all ethics (in the West). It was, or promised, a currency and economy of readabilities and debt – like, one might say, the Greek *autos* or, codified, *anthropos*. He indexes Roman law to facing the accuser and Levinas's claim of face as the premise of Western ethics and identification – even if looking on the face of the deity (presumably anthropoid) is interdicted. Marché's claim is that face as such is dissolved in the shadow zones of trolling into forms of faceless transaction which channel blind aggressions, including rape threats and takedowns of all sorts, baits for the trap of any response or engagement. (The target is advised not to respond, since it could unleash a swarm.) The trolling assaults are "faceless" themselves, operating from invisibilities (or so thought). This epidemic of facelessness appears viral and predisposed within the unfolding of digital and other technologies themselves. It spreads, stain-like, like the black-hooded ISIS to "the West," as from the obverse of a totalized frame it would deface – as, one might say, does Putin's mimetological hybrid and media wars.

This loss of face of the West echoes within, in its Ponzi schemes of credit and bank cartels gaming the global system's phantom debt. Such a supposed "epidemic" could be likened not only to Ebola, memed as a viral and non-anthropic agency parasiting, then transforming cell structures. What is peculiar in Marché's account is that face is mourned not only as if a certain "we" were collectively leaving an entire techno-cognitive (and, of course, "social") epoch. "Face" of course returns us to the core of anthropomorphism, as we think of it, as if "face" were detached and sought in the world of things and forces (like Greek gods): it is bound to any so-called Anthropocene era inescapably and, as such, makes the latter dependent for its definition and ecocidal drive on, among other things, a virtual and now calcified "epistemology of tropes":

> The challenge of our moment is that the face has been at the root of justice and ethics for 2,000 years.... And from the infinity of the face comes the sense of inevitable obligation, the possibility of discourse, the origin of the ethical impulse.... The connection between the face and ethical behavior is one of the exceedingly rare instances in which French phenomenology and contemporary neuroscience coincide in their conclusions.... The connection goes the other way, too. Inability to see a face is, in the most direct way, inability to recognize shared humanity with another....

Without a face, the self can form only with the rejection of all otherness, with a generalized, all-purpose contempt – a contempt that is so vacuous because it is so vague, and so ferocious because it is so vacuous. A world stripped of faces is a world stripped, not merely of ethics, but of the biological and cultural foundations of ethics.... The spirit of facelessness is coming to define the twenty-first century. Facelessness is not a trend; it is a social phase we are entering that we have not yet figured out how to navigate.[22]

It is a "social phase" that nonetheless undoes the definition of the social. Marché cannot yield to a nostalgia of "face," which is irrecoverably dissolved across digital accelerations (and homogenizations). One is now in the imitation of faces, or tags like emoticons. Yet he cannot allow that this welding of face to "intersubjectivity" to Anthropocene 101 ethics may itself have been but one organization among others, a time window like the post-war Western "middle class," or the Little Ice Age, whose passing sprouted Enlightenment memes. Since if face was always an artifact produced by machinal and mimetic triggers, it was always also a technic among others. It, of course, can always be a con, a front for con games and manipulations, Melville's *Confidence Man*. Would the so-called Anthropocene, or at least its twilight today, pretending to come out of the model of the Greek *anthropos* and Roman law in the West, not be in a sense the time of faces? This would seem to resonate with the ritual manner in which something like "anthropomorphism" is casually referenced as a lamentable naivete as if gone beyond – without ever pausing to define or analyze the term itself (*anthropomorphism*) or what it may mean if it all rested, once again, on a robo-trope or metaphoric regime? The problem with "face" is it depends on light hitting it (or the screen), and it depends on the eye seeing it and as if being seen by it – as Hitchcock said to Truffaut, there is "no face," just the way shadows and light coalesce. That is how prosopopeia can be translated, the making of what is before the eyes as face, the giving or emergence of voice in the inanimate, the coalescence of the sense organs of the head in one place, and finally their capture, today, by facial and retinal recognition systems.

Perhaps any fetishization of "a facelessness epidemic" must be questioned for its inverse (or defaced) logic – quite aside from the nostalgia for "face" that is accurately tracked. First, it is a mimetological war that is witnessed, since the mimetic accord that both engendered face (mimicry) and is sustained by it (identification, posters, Leaders, selfies, televisual screens, talking heads). Are we sure "face" has not been a problem – perhaps, if taken as a mimetic regime, technically capturable (telemarketing of all kinds, now "pre-crime"), it is indissociable from a certain, obvious "anthropomorphism" or even the Anthropocene as a hashtag for ecocidal disappearance? Is, or was, "it" one organization of visibilities among others, perhaps itself politically consolidated or used – as when arche-cinema, in Stiegler's insight, both organizes archival communities and perceptual grids, but allows stereo-typal grammars to institute themselves – now totalized today, at the apparent "death" of cinema, its transference from screens to interactive interfaces and neural management? If "facelessness," a turning against the regime of face, is already and preemptively claimed or programmed within digital streams of 0's and 1's, in advance of any cognitive or aesthetico-political turn, one is caught in a deteriorating loop. Now, let us posit a system, today, from which decisions are disseminated and to which all hypertechnologies are subordinate. It takes the strategy of defacement away from any position of resistance, and in fact perpetuates, for now, the ghost ideology of "face," in a streamed, managed and desired mode. Every selfie is a zombie wish for the blood of a self-penetration – which is why I will suggest we look, on occasion, for how a "selfie" of the Anthropocene itself might be posited, marking the lens's and cinema's own agency in what unfolds today (what one may call, in passing, the "cin-anthropocene"). Thus, the "epidemic" Marché marks gives the position of facelessness to the technologies that irradiate rage through the troller, say, from its shadows, unseen (the Ring of Gyges effects), invisible itself (it thinks), dissociated into the dark side of gaming itself (from whose ranks ISIS martyrs and drone pilots are drawn). The "troll" is neutralized as anything more, contained as what presents a symptomology of a wider "present." Hence, Marché's generalization moves from a peripheral harrassment to one that closes out the "we."

If something like ISIS were worth inspecting from the perspective of media and mirroring effects, it yields an asymmetry that returns (literally, today), and haunts the era of climate chaos. It has no pharmacological role to play, no antidote or integration with the heirs of the Crusaders and Isabella. Of course, commandeering refineries and pirated oil undergirded their finances.

Black Suns and Swarming Points

Hypothesis: As we undergo the retirement of legacy literacies and reading transmission — together with the era of the Book, residual attention spans, and migration into screens — the terms of tele-mnemonics and the wars within cinematization arrive as a disgorgement of imaginary interiors, "general semantics," the curation of visibilities. Hitchcock terms and performs this as cinematic and terrestrial "bird wars."

4   Cin-Animation: Notes on Hitchcock's "Bird War"

"Why did he shoot her?" "Watching a ball game on television – his wife changed the channel."
—*The Birds*

Air-design is the technological response to the phenomenological insight that human being-in-the-world is always and without exception present as a modification of "being-in-the-air."
—Peter Sloterdijk, *Terror from the Air*

It was the best of time. It was the worst of time. Was or is. One can say the same, more simply, of time itself among men, which has begun to accelerate into a mode of tempophagy – the auto-consumption of the long-term in the Ponzi scheme of casino referentialism that appears a particular hangover of preemptive digital swarming. The world (the "we" of almost every publication or monograph) has never experienced such wealth, technological perks and media streams – with nanotechnic promises unimagined for the synthetic production of life forms and bodily health (or replacement). Yet the planet itself is a smoking wreck of extincting and unwinding systems and shredded biodependent accords; mutating, it withdraws the complex web of life effects and resources that granted the human parenthesis place and time, and quite quickly. That places our so-called present in a sort of time-bubble in which an irreal set of canceled futures advance and the networks of life that congealed into the "global" present appear permeated by different time-threads (competing and regressed historical models, geological and microbial times, the times of dwindling reserves, extinction events, and so on).

I called this a time of *tempophagy*, when alternative temporalities accelerate a certain effect of auto-consumption. It is a minor category but subtends many others (speed, species-time), and I will explore just a few paths that present themselves. I will add only that I find the crystallizations that occur here neither pessimistic nor worthy of suppression, but catalytic, bracing and interesting. What I will suggest is why, under the circumstances, "climate change" arrives as an affirming and empowering advent – at least from the theoretical point of view, which

is to say, that which separates itself from immediate human concern. One begins, of course, with the full consent to this narrative, that is, to the import of human disappearance.

One might speak without the usual bob and weave that sprouts before impassable eddies – supposing that this is at all apocalyptic. No sudden flash and revelatory phantom of the impact, no suspended now, climate chaos – if we use this superficial tag – is itself banal, having to do with the slow-drip of times, the wearing away of coasts and compacts, "population culling" spanning decades, the daily haunt of interminable heat beyond the usual century of speculation? If "climate change" is used to designate an entire dossier of mutations revolving about carbon release (the transfusion of dead, terrestrial organisms over millennia into energy released into the atmosphere), it is both verbally redundant and mis-literalized. Today the self-feeding backloops and processes underway traverse oil consumption to agricultural depletion, extinction events, the transference of uncountable chemicals into human populations, the advance of technologies and splitting of castes on a proto-speciesist level – and the unrepresentable horizons of a mutation. And we are already in the middle of it, as reports from numerous outlying zones of the imaginary global empire report, to say nothing of the nth wave of mass wealth shifts and consolidations at the remote top. One witnesses the low end of this in the Trumposphere's will to shred health insurance and heap debt on Trump's suckered "base" and a once-middle class. The latter produces different symptoms: religious regression (linking evolutionary denial to that of "climate change"), activists for humans extincting themselves before all life systems on earth are compromised, techno-parasite phantasias of singularities, evolving enhanced cognition and replaceable organs as prelude to planetary escape.

One might replace the term or phrase "climate change," which was engineered for failure, with accelerated ecocide or global mutation. The word "change" is too aimless (what doesn't change?), pseudo-mimetic, passive; and "climate change" thought locally is tautological. Climate unnames what one is wholly embedded in, without distance from, air and currents, food-chain circuits (and media circuits today); like a fish in water, the occupant is never in a position, quite, to stand apart from what, too, involves his or her or its embodiment. "Mutation" might apply better – the simple naming of a technogenic, biotic mutation, a mass extincting that appears autogenic and suicidal. It operates as if an interruption, without being so. Below, I will read this problematic as at stake in the spellbound polis of Hitchcock's Bodega Bay – as a name, too, for the new public space that the screen itself has become. The zombie chant of the schoolchildren ("Now, now, now") mimes the paradox of a tempophagy that formalizes itself as a mnemonic repetition by a progeny awaiting to be given their memory tapes. All of this before the swarm attack of an avenging cloud of flying marks that disassemble the visible and dispossess the very concept of the home, the *oikos*, the interior, with a figure of animation irreducible to represented animals ("birds") or technics alone. I will suggest that this nanoswarm which empties eyes and interiors arrives from a position as if turned against humans, against "the human," from a position of cin-animation that bespeaks a "life" effect which, like the former, is neither of nature or an Earth as such. In the pointillistic attack of the bird swarms, however, which displace and then shatter into multiplicities any sun as source of light, these technic "points" manifest a digital logic which the analogic techniques of twentieth-century cinema would, for Hitchcock, harbor.

■ ■ ■

Representations of "climate" catastrophes tend to mask an impasse in representation. Amitav Ghosh platforms his calling out of "the great derangement" – specifically, that Asia had stepped into the Western hyper-industrial trap, sealing the ecocidal *vortex*, against a complaint that literature, too, had been in a mode of de-nihilism by not writing to this horizon, well or at all. That could not be said for cli-fi cinema, and cinema might be deemed to overlap with the forms of biocide underway. Such arises not because the latter lies outside of humans and have no face to identify with or against, but because the figure of "climate" appears encompassing, non-personifiable, processual, of no precise time. It also generates the suspicion that "representation" itself participates in its acceleration. Such would include conventions of reference and the visible inseparably linked not only to semantics, or the home, but to interiorization, consumption, eating, or a certain anthropic blindness of what might be called, today, less the "hyperindustrial" than simply the late "Anthropocene era."

Climate activists have been advised that too much media information risks producing not response but numb passivity. This casts our artefacted "present" in a counter-Hamlet position. It knows too much from a (future) ghost, yet the knowledge does not coincide with phenomenality or experience before it in the court. It is too much – and, today, one finds new temporalities unleashed. Like the zombie banks, zombie media and, often enough, zombie theory today, the cognitive "present" in this regard seems to reflect on its own necrescence with fascination – deferred referential Ponzi schemes of debt and vandalized "futures"; it seems broadly unable to imagine different variants of life than those familiar, triggering, broad and intensive relapses from Obama's presidency to "sustainability" tropes and neo-humanist semantics. Critical force here might be thought to shift from a focus on the otherness of the (human) other to something "wholly other," from the human-on-human model of struggle to the "threat without (human or visible) enemy." Nor should one pretend this conforms to apocalyptic tropologies, wishes or linguistic rushes. What names or inspires these horizons is the opposite of "apocalyptic," doesn't gather itself in instants, flashes or revelation precisely. Without face or personification, it ("climate chaos," if that is a passing basket phrase) encompasses a biomorphism in slow motion, irreversible, non-linear, unutterably banal for involving molecular, dynamic, chemical, elemental accords. In its way, for men {man?}, this locates itself in the "unthinkable," residing outside representational, mnemonic, conceptual and cultural premises of twentieth-century critical projects. It arrives in the form of glacier meltoff, or dying seas, or biodiversity collapse. It links the artefacted and screen-dwelling "present" of today, in a time-bubble, to prehistorical and zoomorphic timelines.

I

There will likely emerge a critical discourse of catastrophics in the twenty-first century to incorporate the archipelago of mutations now calculated under the general imaginary of climate change. Such logics encompass various spinoffs and feedback loops (mass extinction events, resource wars, nanomutations). If the category of "the catastrophe" is temporally mobile, it is unlike "trauma" in being neither a faux origin of memory nor subject to allegory as we know it – which turns on the formalization of anteriorities, ruins. These incursions on a planetary or biomorphic scale do not contribute to or further current critical agendas, cultural identity, social justice, human-on-human narratives. Like Hitchcock's birds, they appear as if acausal.

Among the images that struck me of a certain "materiality" was one from New Orleans – in the parenthesis following its deluge. Cut off from communications for days, people faced a soup of contaminants, a site where techno-poisons and a prehistorial swamp reversion opened a temporal warp. A pause of non-response and abandon, no tidal movement. Yet as an entry into the dossier of catastrophes, it is also an image of the state of the image, the seepage of the non-anthropomorphic into the spellbound frame.

New Orleans suggested a lateral acceleration from the staged shock of "9/11" – only without human double or aura, without face. Leave aside the subsequent militarization of the gulf zone in a cleanup peopled by troops back from Iraq for whom it looks worse, reportedly – the ban on showing corpses and so on. Leave aside the triage of a disempowered class, the racial subtext or the dubious reality TV that elicits a strange voyeurism. What interests here is the rift in the anthropomorphic bubble, temporal and perceptual, the breakdown of circuitry in which *terra* substitutes for the spectral double of the terrorist. Hitchcock casts this assault as the "bird war" of the present – that is, it is a "war" that turns against the two sides of so-called world wars, hot or cold. It must, however, be thought cinematically, also as an experience of and within a cinematic "modernity" in which memory programs, technic streams, telecratic spells and the virtual bubble of a certain *telepolis* are assumed – that is, it is something that happens of and within not a cinema that represents some real but an already cinematized psychopower captured by the instrument, here, as if turned against that (and itself).

Hitchcock sees the world wars as late imperial contretemps between two extremes in the same Enlightenment epistemology. It opens and supersedes a certain "now" – call it, the era of climate change, mass extinction events, the shift from an "otherness of the (social) other" to the non-anthropomorphic. The schoolchildren in their mnemonic chat conclude, again and again, with the repetitive assertion, "now, now, now" – the repetition by the zombie children undoes the immediacy it would somehow stamp. It is absolutely anapocalyptic, this "now, now, now" that would isolate a singsong present in the aesthetic rehearsal of a memorized chant. It is one signature of tempophagy, the very instant aimed at assertion, highlighting its increasing formalization and anaesthetic urgency. The birds assault the "eye" in its blind programming, in the name of its cinematic other – and against the cinematic accelerations that produce a spell.[23]

As such, the birds are not *apocalyptic*, as Lee Edelman instructs us.²⁴ They reveal nothing; they do no end "time"; they dispossess and blind blindness. Their invasion as a warping of temporal logic implies a folding in of the frame, without outside. Nature is the effect of certain semantic or mimetic practices, and its disarticulation accompanies that of mimetic regimes. What may seem parallel is the nausea of Zarathustra before the backloop of eternal recurrence (Cathy's vomit), and the biting off of the snake's head, the stepping "beyond" a closed circuit to a site that is not mapped yet in other terms. These birds are not animals but technemes, allied to machines, telegraph wires, pecks, the prehistorial and post-anthropomorphic, the cinematic as such. Earth, after all, is what the camera will always (also) be gazing at, be recording – even when a simulacrum "Earth," as with sets of anthropomorphic monuments made of stone and steel. But nature was always other than maternal, an anthropomorphized "she," the shot's representational claim. "She" is something else, proactively mimetic, a mimesis without model or copy, much like species which alter ceaselessly according to the technicity of an environs or for camouflage or shape shifting, adapting proleptically as an animal or coral sea creature or insect assumes camouflage before a predatory other when it cannot "see" itself to be like the mimicked twig or rock or leaf. This artefactual "Earth" is the paradoxical counter-world to the passivity of "globalization." What may be called this *aterra* is allied to the birds.

In this twenty-first-century moment, decisions made by the present, the logic runs, alter or erase prospective futures in exponentially calculable ways (tipping points pass, glaciers melt, coasts inundate, "population culling" occurs). Hence, the appeal to children, or grandchildren, like a recent commercial featuring a train which a man steps aside from, revealing a granddaughter in the way – generic and without traction, since there is no ethical contract to what does not exist yet.²⁵ There is no general ethics of, or toward, virtual "futures" – particularly ones not appealing to contemplate. Such a global "present," in a sense, hoards or consumes temporalities in a sort of tempophagy, as it does life forms, species, resources, and so on. Hitchcock names this scenario "Bodega Bay" and, when all is said and done, what is attacked in the name of these birds appears in one register to be how the visual and the eye are programmed, with memory, in the mode of blindness, commodified psychism (faux Oedipal scripts), consumption. I ask below whether Bodega Bay's "bird war" may be read – an uprising of the trace itself from within the zoopolitics of the image – as a parable of this incursion. For such an autoscopy of cinema's zoopolitics, one must turn not to the genre of disaster (or ecodisaster films) now in circulation – which are bound to representational fantasies, apocalyptics, and inevitable returns to the family, survivor, of the home essentially – but to where the logics of cinema, fully involved in techno-wars and species eviscerations, in spellbound psychotropies and memory management, asks this question of itself. The logic that issues forth is not so much exemplified by driving the Brenners from the home at the end but by a foreclosure of the metaphorics of the home, a dispossession of the *oikos* (or ecological), and a mode of sight that is, from the birds' perspective (with which Hitchcock identifies), blind.

II

As a hyper-medium that absorbed others, "cinema" marks itself as a historial event producing new forms of memory and perceptual blinds. In Hitchcock, who can be read as a courier and critic for this advent, there appears a sort of Benjaminian war over two logics or definitions of the image itself: that is, the technologies of memory. This war can appear as that of the home state, or rote memory and mimetic spells, and that of the cinematic saboteur, or assassin, whose uprisings, if successful, would recast the MacGuffin of history and the senses. By late Hitchcock, this war shifts from being between human communities and appears as a "bird war" against the human as a construct – and against the latter's ocularcentric programs. These birds, black slashes or wingbeats that peck out eyes, are a sheer technicity. They avenge from a prehistorial logic that, simultaneously, anticipates a coming "war" over these inscriptions themselves – what might be called a "war within the archive: from which phenomenality and programs of consumption are legislated. The accord between non-auratic cinema and Hitchcock's attacking Zarathustran birds poses a biotechnical question: what role, in the post-global orders of today, does the cinematic image have in the ordering of the "senses," the contemporary spell of political anaesthesia, and the mass manipulation of memory? What is implied by the paradox of a cinematic practice that marks its aim not as servicing a visualist culture but as violating, blinding or dispossessing "the eye" as an artefactual construct? The "bird war" in Hitchcock brings the figures of animation and the house into collision, implying coming wars of reinscription of the senses that open upon a post-global politics of memory.

How might one identify these slashes that carve up the visible, dispossess the house, empty interiors, put out a certain way of seeing? We know, there is no time in which to interpret this attack – nor is any single referent to hang on these animated points allowed to evoke animation, the prehistorial and the mark. Critics generally analyze the misfired scripts and psyches of the characters, grouped nonetheless without star power or narrative cohesion. How to address the birds themselves, which Hitchcock alludes to as the film's "stars" – shifting identification from the B-list humans, squeezed finally from the frame, to their atomizing antagonists? We can here slide into a remark about Marnie: "No references," is first recalled about the blond thief when arriving at her first workplace, Strutt's, before its vault is emptied. From this perspective, they perform an assault on "general semantics" – which Melanie is said to study at Berkeley – from a hypothetical outside that is, nonetheless, the mere points or marks out of which the visual and the screen coalesce. These small cuts or points are reanimated, militarized, rise up from the margins, drive the inhabitants out of the frame. The irreducible mark embedded in all visual constructs arises to avenge itself, or wipe clean, a sensorial program bound to hyperconsumption, anthropomorphic blinds, terrestrial eviscerations. Hitchcock never recovers from the experiment that outbids cinema with itself, turns the latter against itself suicidally, realizing the bombed bus of *Sabotage* on-screen, in a logic of sheer exteriorization.

To begin with, one must assume the action of this work occurs on the screen, too, precisely when these prosthetic slashes carve up the title letters or the visible itself. Yet Hitchcock's birds avenge in the name of a certain justice, presented in the film as an absent cause. They peck out eyes – and assault a certain way of seeing, which is also to say, here, of eating or interiorizing. Tippi Hedren's platinum head and mannequin looks (stepped out of a television commercial) violates the visible – turning the blond as faux figure of metaphysics into its other (to become the dissembling "Marnie" as site of truth). Melanined, cohabited by a figure of blackness, both white and black are consumed by the same Heraclitean semioclasm – become "one." Bodega Bay's "packaged goods" recast the humans as bad copies, replicants from a B-cast, or real "models" inducted, like Hedren, from TV advertisement logics into the frame – a precession of the simulacrum by itself, indicating a rupture, triggering the attack (as the hysterical mother in The Tides restaurant misrecognizes). In driving the children from their moment of memorization, these Zarathustran birds posit an event or possibility of ex-scription. It evokes nausea. Why?

Technically, since the screen is "Bodega Bay," the birds simulate innumerable points (as Mitch Brenner mumbles, "What's the point?"). Their swarm precedes any graphic image that might coalesce as animation on the screen – any face, but also any alphabetic sign, or hieroglyphic. Cinema in this sense precedes and supersedes the era of the book, which is also marked in Bodega Bay variably – from the post office to the schoolhouse. The schoolteacher Annie's line about herself, spoken before a library cabinet, addresses this when she says:

"I am an open book, or rather a closed one." Open: that means, the histories of the book are all evoked and traversed in this site, return at the atomizing assault of marks, and this by what precedes letteration itself (and the histories of alphabetic writing, hence monotheism); closed: that means, with this release of the cinematic mark, the era of the book is closed, dissolved. There seems here a movement beyond twentieth-century epistemo-critical practices – the acute focus on the social other, the "otherness of the other," identitarianisms of all sort, historicisms and dominations, so-called ethics – and a shift toward what cannot quite be called the "wholly other" represented by the attacking birds. No Enlightenment, dialectical, utopist, materialist, faux nomadic or psychist discourse seems adapted for what alters the contract of temporalities and biopower. "Absolutely No Credit," says a sign in *The Tides* restaurant.

One might at this point riff on the following here: the first sign on the cinematic trolley car advertising "The *bar* at the top of The Mark" (quote from *Vertigo* – drawing our attention to bars and marks at once). To be without "references" is not to be without violence, only to remark that the way perception is programmed, reference regimented, may be structurally blind, consuming. The "birds" generate reference to account for themselves, only to exceed it as cutting wings and myriad attacking points. Why?

III

Interiorities of all forms will be vacated or exposed as mere pockets or folds like the bay itself. Eyes are pecked out, stomachs emptied, the premise of the *oikos* abandoned. Nausea sustains its logics or is passed through by them. And this is accomplished by a nanoswarm, disciplined yet relentless, which is experienced on the screen as so many points, irreducible cuts and marks preoriginary to any era of the book, to alphabetics or hieroglyphs.

The logic of attacking birds is that of so-called cinematic "shock" carried to a formal extreme from which it inverts as normative – the aerial bird shot. If we view "Hitchcock" as an unreadable event within the prehistory of contemporary teletechnic culture, one across which the formal thinking of script and cin-animation reflects on its destinies, certain moments or turns put the histories of the senses, memory and the machine into question. In such sites, wars within the image from out of which the "global" would be constructed are underway. And one finds reading models that exceed and precede old sensorial programs like ocularcentrism – the auratic programming of identification and mimeticism that remains the official ideology of the photographic

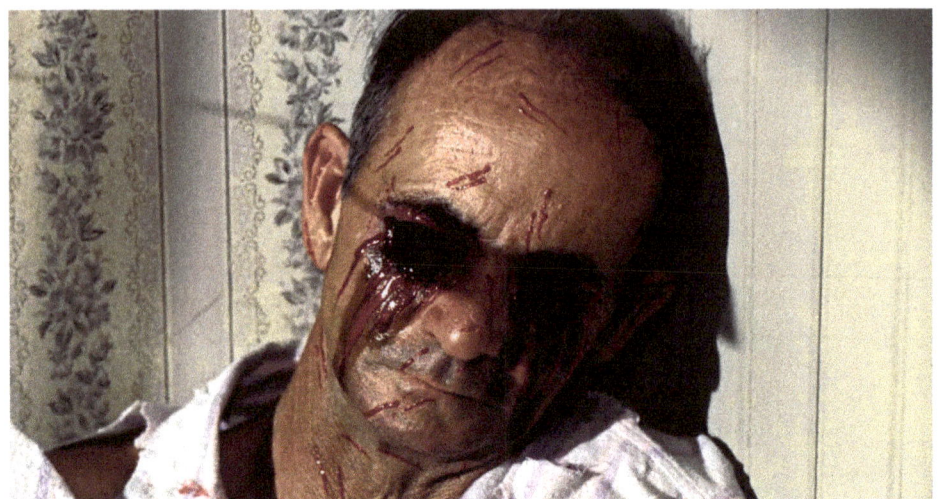

image. The birds, for instance, issues forth black, slashing wings that carve up the visible and peck out eyes from a non-anthropomorphic, and perhaps avenging, prehistorial space. They gather for an attack on a *jungle gym* before a school house – when they are not flying past telegraph wires or attacking telephone booths, allied with machinal hums and vehicles.

Melanie's past trauma involves the shattered glass pane of media and what is obscurely called the playtime of her summer at "Rome" – that is, everything ever meant by "Rome," as *Spellbound* reminds us. Rome: that is, the capital or head of the empire and the church, rewritten as the streaming lights of the cinematic apparatus, programmer of memories and new technology of the spirit (the movie house as inheritor of the theatrical-churchly ritual). Again, the word "capitol" will appear on a sign at the gas station when attacked and in flames: "Capitol Oil." Oil, ink-like, allied to the head or *capo*. Engine of the teletechnic era and its vehicles of transport, it yet circles back across aeons, tapped by man as energy, become the electric bulb of cinema, inhabiting "light" itself. The oil is allied to these birds that, the progeny of dinosaurs, return from above – to set a gas station ablaze, set off fire alarms in the archive.[26] Without references, the birds solicit and disown in turn every interpretive program – yet, conversely, are identified with, if anything, what precedes the letter, the voice, "mother" as such, the unearthly earth and fluorescent horizons where sea meets earth and then sky as so many lines.

Why do these birds attack the eye (the most frequent *word* in the film is "see")? Why are they identified not with animals and certainly not with an avenging "nature" (as Hitchcock pretends to Truffaut and in his trailers) but with engines and machines, telegraph wires and telephones, spatial graphs like the jungle gym? Why is Hedren linked to studying a course in "General Semantics," and why does the name Melanie, a black name, cast her platinum-white screech as a blackout, as though white and black collapsed, momentarily, as the same Heraclitean semioclasm, exposing "light" itself as other than itself? The identification of the skeletal graph of the jungle gym is with the very edge, the rim of the archive or great house, lines and joints, vertical and horizontal lines, cells, frames, rows or bands – as if the avenging black flecks, attacking eyes and children, outside and against the "human" as such, were the uprising of the frame folding in against the MacGuffin of an imaginary content. So when one is looking for referents or tropes to rename these figures, one cannot call them "animals," or "nature," or "apocalyptic agents." If one continued this prescribed game into the present and asked them to conform to an allegory of climate change, as I am doing, they would resist differently – being allied first of all with marks, machines, telegraphics, technics.[27] They would stare back.

The birds are, the ornithologist Mrs. Bundy says, "impossible," the *impossible* – they, these marks, were not supposed to violate the logics of the visible, which they, after all, gave rise to and serviced. Zarathustran harpies, avenging on behalf of a mnemonic backloop that always precedes the screen present it generates, they turn against the very construction of the "eye" – and in the name of an other that is not just without identification, not only prefigural, not only a convergence of technics and the province of animation. Preceding and closing the era of the book, the birds return from the archival rim or dome. It is interesting that the site beyond relapse, corresponding to leaving the house, abandonment, seems represented by the mute, catatonic, emptied, staring eyes of Melanie. This stare, which appears elsewhere in Hitchcock as a zero-figure or so-called psychotic logic, is actually a cinematic *apsychism* broadly – Gregory Peck's amnesiac trance with a razor in *Spellbound*, Norman at the bog, are here assigned a future. It exceeds any psychologizing trope. When Lydia rushes from seeing the blinded eyes of Dan Fawcett, a summary of all literary blindings of the blind, she understands too that no one knows where "mother" is (as Melanie says), that she (Lydia) is from the start a simulacrum mother, a copy like Melanie whose hair rhymes with hers, and that one proceeds into Bodega Bay's hermeneutic traps without any familial, Oedipal or "maternal superego" map (as Annie more or less tells us: "with all due respect to Oedipus, I don't think so"). The characters are largely ciphers and not personalities: interpretive vortexes scanned against what may seem to be sheer exteriority. Melanie, on the buff, notes that she doesn't know where (or if) *mother* is.

IV

We are told several times of Melanie's obscure trauma in Rome, of her Rome days, which she has left behind her. Something about reckless play, naked romps in fountains that made it into the newspapers. Something about a wild past linked, less obviously, to the shattered windowpane of media which brought her to the courtroom where Mitch first saw her. What Hitchcock puts in the past is the entire Christo-imperial legacy he evokes elsewhere in references to "Rome" or "empire."

In *Spellbound*, "Rome" is relocated, just after the Second "World War," as New York City, the new mediatric capital of the coming global era. And we see a banner above Grand Central Station – that is, the sort of historical ground zero of the arrival and departure of all historical trace-chains, cinematic times, tropological narratives going back through the archaeological capital of the old empire and the Church, of the Europe or West decimated after that war. That is, before its coming amnesia in the era of the global announced by moving the capital to New York City, new capital of media and advertisement, the Empire State. At the center of the station vault, the banner advises the ticket-buyers below: "Buy *More* War Bonds." *Grand Central* appears, then, as a passing allegorization of the screen itself, that very one before us. From the metaphorics of cars and planes and ships, from Virilio's dromology, that movement arrives at the station of arrivals and departures, Grand Central. This new public space, which is the screen itself, mimes a ground zero of stationary transport to all possible names and places, a central station that is the screen. The name "empire" appears as another sign, a rectangle mirroring the movement of mere traffic. The Empire State Hotel glosses this global empire to come, allied to cinematic logics or accelerations, and it is here that Hitchcock's cameo appears.

The shot of Grand Central prefigures Bodega Bay. It streams with cinematized light rays from its giant, eye-like window, vaulted like a cathedral, transposing the old Rome of empire, cognition and the Church into the implications of an artificed technology of the spirit (and memory, or amnesia) dependent on machines of transport. The cinematic bars echo those that haunt Peck's amnesiac. The non-site of arrivals and departures, a giant, domed head, is where memory loops are installed and spells reign. Throughout *Spellbound*, there is a waiting for what is called the "new head," who never arrives, or already has as an imposter. The war is over and seen beyond by a weak, messianic urge undermined in advance. Thus, one should look ahead and "Buy *More* War Bonds." The bonds in question, echoed in the titular

Spellbound, would be between cinema and psychoanalysis, or both and "the war" (or war as such), but it also tells the ticket buyers below that the war is not over, that the catastrophe of Western history just witnessed is already subject to amnesia, that the contretemps of fascists and western, liberal, colonial democracies was as if between two extreme doubles in the same Enlightenment episteme, a late imperialist and territorial skirmish. That is, that the real war to come, that over the Earth or its prosthesis (an Earth without nature, consumed and cinematized, without future), is announced already by the "global" logic cinema's arrival marks in advance. *Buy More War Bonds*, the giant placard urges above Grand Central Station – where, like Rome, albeit cinematically, all tracks lead. It anticipates another war announced or to come at the amnesiac close of the World War itself. What will be called the "bird war."

V

How does one strike against a totalization one cannot, per definition, see – particularly one that, as one perhaps witnesses in post-democratic "America" today, a land of telecratic denial and suicidal trances?

To launch a war on what is essentially the visible occurs in the name of what also can be said to make that itself as if possible. Or in the name simply of the "non-anthropomorphic," an assault on temporalities and sensorial programming, an intervention in and against programs of memory as such. The birds correspond to flying marks. As cinematic black holes, they withdraw from personification or anthropomorphism – no aura, no star power, no metaphoric viability. The birds avenge in the name of a nameless techne outside of the spell of ocularcentric programs, in which the eye is that of the hunter, the reader, the consumer, the eater. Gathering on the structure of the frame, no longer house or interior, allied to memory and the disruption of mnemonic bands, they totalize and dismiss a form of terror. It is an assault against an artefactualization of "life" or sight referenced to global markets (Bodega Bay) in and by the cinematic operations that have, over time, produced that relapsed regime.

Does this "bird war" from what is the inhuman and the technogenesis of animation as a form of backlooping, wild a-semiosis, suggest still a transvaluative logic? In the name of what does Hitchcock so clearly side with the birds' rigorous justice? This is clear enough at the end: the dispossession of the house and exodus of the Brenners' revving coupe. How does the MacGuffin of a cinematic uprising that is routinely suppressed at border crossings in the espionage films, the model of early Hitchcock, resolve into a "bird war" against a version of the "human" and the screen itself in toto? Why the Zarathustran and exploding backloop to these black nothings, flying marks and black suns which induce nausea and vomiting in little Cathy? Does Hitchcock's term for the cinematic variant of eternal recurrence, that is, "vertigo," simply come to this – the sheer exteriority of all, the folds and pockets of the Bodega Bay coast, Ekel or what exceeds it?

Hitchcock sends us back to the logics of *Sabotage* in the opening visit to the bird shop, where Melanie asks for Mynah birds – birds that memorize human language as sounds and repeat phrases with no subject to the repetitions. Later, when the schoolhouse is attacked, the children driven out when they are all repeating songs and phrases together, like the Mynah birds sought but missing in the bookstore – aligning "human" memory with such group imprinting – mimed by the kids group displays of upset when told they will leave in a fire drill as the birds gather to attack. The cinematic assault occurs where memory is inscribed, programs of reference memorized or *recorded*. The reference to *Sabotage* is the bird shop – which is a front for the bombmaker Chatman in that film, who supplies the bombs Verloc will issue from behind his front as proprietor of a movie house. The bomb is allied to a bird cage and the explosive "blast" of cinematic shock to the successive bars – the series of alternating slashes that precedes any sign system and occurs in every Hitchcock work as a signature effect – and to the moment the birds will "sing." It would be placed in Piccadilly Circus, what is called "the center of the world" – meaning where phenomenality is set, the spectralization of worlding governed. (*Circus* retains the reel-like mechanics of a circle and circuit, while "Piccadilly" incorporates Hitchcock's favored letteral rebuses – the Pi, the CA (31/13), the initializing "bar series" (// or ll).)Cinema, with its memory band, mimes what de Man called the "phenomenality of inscription" – and it is by atomizing these that a moment of disinscription is approached. "Circus" inscribed the circularity of a temporal backloop that the mnemonic band mimes and forgets in advance of itself. Yet there is a zoopolitical dimension to this act – as it is called in every performative sense in the dialogue. This is explicit in a visit to the zoo to visit Verloc's handler, who is not impressed with his mere turning off of the lights in London – a caesura and blackout, displaying "light" itself as a techne. They didn't get it. The Londoners coming from the cinematic underworld laughed, thought it more entertainment. The ante must be upped. Yet in the aquarium, it is against a screen-like tank that the melting of building and structures is projected – as if this atomization occurred before and in the name of prehistorial and premammalian sea creatures. Like the bizarre fish that a passing couple note changes sexes after laying innumerable eggs – a shape and gender shifting life forms, as though "life" were itself, here, the effect of a complexly mutating technics. Like animation – which supplants the mirage of "life" with what is not alive but itself generated from cinematic graphics, a mimesis without model and copy mutating forward. What is marked is a technic "life" that stands beyond the artefacted binary of *bios* and *zoe*, outside sex difference or reproduction (Verloc's sexless marriage, Sylvia Sydney's sailor-boy clothes, the boy Stevie obliterated, fish that change sex, and so on).

Yet this is what is shown the one time we visit the theater itself, where a Disney cartoon of bird-men and a bird-woman appear, the latter as a Mae West figure – what is herself marked, in Hitchcock, as a female *female* impersonator. When "the birds sing" turns out to be this crooning of half-animeme, half-human figures, the bomb affiliated not just with the trope of explosion, of blasting and shock, of atomization and artificial memory, but the "secret" that the cinematic precedes and generates what takes itself for "life," or "the home," or the "human" for that matter. It is an artefact and artifiction – together with the concept of the "present" we inhabit. It is, it accesses, other temporalities, as the time-bomb makes clear – which explodes, too, a prison of temporal definition, anthropomorphic. The cinematic allies itself here, again, with the prehistorial and preoriginary, with an animation that dislocates the field of "life" into other terms – as another effect of technicities and artefacted mnemonics. Yet here there are two logics at war: that of the house, of mimetic cinema and anaesthetization, the spell of ocularcentrism and the hyperconsumption of the planetary – that is, the image as we know it, a blind construct – and its other, the cinematic assault of the birds. The bird attack is one on an epistemo-anthropomorphic regime by its own premises and techne, an assault on a way phenomena and the visible had become programmed, packaged goods.

VI

Hence, farmer Dan Fawcett, the blinded man – who is shown with eyes pecked out. What is assaulted by the cinematic occurs at or before the core of the visible, the fiction of the consuming "eye" (whose logic Hitchcock allies to that of advertisement media) in the name of alloanthropomorphic legibilities at once teletechnic and preoriginary to four thousand years of scriptive history and animal forms. Oil, engine of the teletechnic era and its vehicles of transport, is allied to these birds – {already mentioned above} fire alarms in the archival order. This non-origin is, was, in the preceding film, called "mother," a figure everywhere yet without gender or one "voice," without place, a *khora*-site in a Derridean sense, a non-site of inscription precedent to *phainesthai* in its entirety.[28]

The schoolhouse is spellbound – like the last man of the mediatric era, the ocularist consumer of democratic fictions in a post-democratic horizon for a while sustained by a "global" war on terror without face or temporal or geographical horizon. The burning horizons of Bodega Bay, all but electric, shift the problem of the so-called living and the animeme into one of animation and trace-chains that revoke any definition of earth that relies on maternal metaphors of ground. It is not the "natural" that attacks. One could call this *Aterra* – and it demands a certain vengeance against the human program or artifice by what can be called an "aesthetic or epistemo-political agency." Hitchcock saw the "world wars" of the last century as fratricidal contretemps played out by opposite poles in the same Enlightenment template, and anticipating (or tracking) the coming wars that would issue from, and against, the subsequent totalization of "empire" – which he located in New York City.

To inspect any zoopolitics of the image, its many spells and blinds, what it frames and anaesthetizes, is to question the post-democratic era we have stepped, as if spellbound, into. One can wonder whether the so-called global war on terror was connected to this media totalization of an aesthetic ideology. It gives rise to a spectral war that is supposedly totalized and against another without face, that is without temporal or geographic borders, and that occludes and diverts seeing the greater threat to the "homeland" from the machinery within the accelerated systems itself – the predicted decimations of climate change, bio-pathogenic collapse, oil and water wars, and so on, whatever lies outside what Agamben calls "the anthropological machine." These neutralizing logics of advertising and consumer identification shield the emergence of what one might call, inverting a Derridean fable, the "autocracies to come" (as the little girl Cathy says: "Mom, I know all that democracy jazz"). One summons, unleashes these birds – "Hitchcock" does, but I am thinking of ourselves now – *not* in a state of emergency, but rather after that had itself been anaestheticized, totalized without horizon.

The birds attack the construct of the visible, standing reserves of reference and consumption, interiors, eyes since neither the marks on the screen nor the reading audience that has entered its compact is removed from this circuitry. One witnesses a sort of suicide of cinematics in the assertion of a violence, and violent reset, as if excluded from the visible itself. The interruption of memory programs explains the birds' fondness for cutting down children. Bodega Bay locates a vacant yet consuming pocket, that of hiatus or standstill, *at(r)opos*, the market bodega a normatization of cinematic "shock." Like the catatonic Melanie who departs the film in the wounded image of the speechless, the zero, the wiped clean, the screen-like "consciousness" proceeds in a post-traumatic and asubjectal reorganization.

"All the windows are broken in Dan [Fawcett]'s bedroom," says Lydia to Melanie, "all the windows." What occurs when the media glass shatters, or becomes visible as a spiderweb of crack lines drawing attention to itself when a bird strikes? The nanoswarm of these wingbeats and zero-effects inverts the logics of the MacGuffin and, with that, anthropomorphism (the birdsas "stars"). One is asked to negotiate what precedes letters and the spell of spelling itself, of getting one's name in print (as is said of Melanie's scandals). A chiasmic X appears throughout Hitchcock. Taking on different forms, it mutates, interestingly, in the late work. From the two Charlies of *Shadow of a Doubt* to the "crisscross" of positions by Guy and Bruno in *Strangers on a Train*, to the "imminent Dr. X" that the amnesiac Peck calls himself – linking a certain cure to this cinematic chiasmus or void. It is seen on the back of the servant Germaine in *To Catch a Thief*, a servant of the Underground ("she strangled a German general once – without a sound"). Then, as if tiring of its automatism and ubiquity, Hitchcock targets it with the single assassination shot of the second *Man Who Knew Too Much*, as the flag beneath the Hitchcock-like Prime Minister suggests (bald, rotund, infantine), as if it itself could be exceeded. And it then turns up written onto the face of the earth itself in *North by Northwest*, a giant chiasmus presented by the crossroads as the prairie stop, as if between the earth and the living. But in *The Birds*, an "X" emerges on the screen in curious places. One appears on a neon sign for Lucky X Lager on the window of The Tides, for instance, where the community goes to imbibe and feed, and then it appears on the car horn that Melanie frantically strikes in her besieged Ford Galaxie.

The chiasmic X is here stuck in an irreversible limit, its horn blasting. It no longer reverses character position or polarities within the visual but whatever is within the frame with something else. It may be that the twenty-first-century critical project will be to attend to an epistemological mutation not only of memory regimes and referential orders but inscriptions that organize perception and the ritual of thought.

VII

What emerges in this excess of cinema, turned against itself (or the eye it generated and serviced), turned as if suicidally? Do the massed birds at the end resemble a frozen swarm of black phoenixes at the edge of the visible – or the archive as such? The exodus from the *oikos* that the Brenners' disappearing coupe suggests is not an exile, the romantic leaving of a home one can recall (like an Earth as it was), not a diaspora, since there was no home to return to. The birds will now be everywhere else, inescapable, the air itself, moreover in "light" itself – irreversibly.

The screen marked as the new public space becomes the polis called Bodega Bay ("our little Hamlet"). That is, the apolis or mediapolis whose circuitry constitutes "public space" without place, the polis without polity. But what seems ruptured too is the biopolitical model itself, which appears now as yet another enclosure or anthropic ruse (the last nihilist avatar of humanism, say), as if from a without. The nameless sight of the *zoe* does not deliver natural animation or even food. The model itself, which has the *bios* in control of the (included) excluded *zoe*, is now enfolded by what includes both as if within its dossier. The birds enfold an entire biopolitical regime into themselves, like the invaginated fold of the bay. They swarm it, surround it, peck out its eyes, eject it from interiors. And this, not in the service of some name (the animal, nature, technics itself) but of what precedes letteration ("General Semantics") and performs a vision of techno-animation as "life." Theycannot be distanced from the memory loop that generates the screen itself. From its perspective, from that of the oil or the bird's eye, that is, outside personification, animation appears otherwise. Cinema here marks and separates itself from the human community of Bodega Bay (or the screen), who are no longer the "stars" and are driven off. In the "bird war" cinema marks its own awareness as an agent in the mediacratizations of techno-war and propaganda, representation and hyperconsumptions and speed and here separates itself from those effects and histories, as if it had all along known their suicidal direction.

The *Mae West* Disney cartoon which appears centrally in Verloc's movie house, the scheming center of *Sabotage*, crosses a human with a bird. This would have been, in a sense, Hitchcock's figure for "life" as animation, without one species or gender. Such a logic (which cinema would have to claim) displays a vision of an Earth and its life forms, held by the visible, as thoroughly proto-technic, without "nature" or "mother," caught in the backloops of mnemonic programs and machines – whether that of DNA or modes of perception.

One may situate the era of cinema within this criminality or blind portrayed in *The Birds* as that of telecratic "humanity." If so, might it manifest its alliance with the zoographic, animation, the *animeme*? If programs of representation fueled, unaware, suicidal accelerations, could cinema not in resistance turn against itself, appear suicidal, sabotage or attack the eye (as if in the name of a justice)? Does it wield "justice"? Does the screen itself gaze back, revoke the personifications it elicits, turn against the consumer of its product? Is it not itself in fact prehistorial, structurally zoopolitical or, more precisely, outside the figuration of "life" it generates as animation? Usurping the power of the human B-listers, it recasts "life" as animation – an effect of cin-animation – which includes the technogenesis of life forms produced out of generative mnemonics (DNA, cellular coding). The birds are tied to marks, technics, measure, shifting biomass, assaulting non-linearities and cascading backloops that arrive as if from "outside." This complex without a name or recognition factor is outside man, the "other," animation as such. From Poe and Mallarmé, we know they are identified with a midnight hour, a proleptic suicide as cognitive and systemic mutation. The perspective of *cin-animation* is destructive of the human construct – its blind consumptive use of the eye (linked, still, to prehistorial hunting needs, transferred to the "capture" effect of image and the neural sublimations of reading). Hence, the closed book that is, at the same time, open.

II. HAVOC IN THE "PLANTATIONCENE": ESCAPED SLAVES

On Dis-Inscription — or, the "Junctureless Backloop of Time's Trepan"

Hypothesis: The prefigural form or mode of blackness has all along absorbed and pre-inhabited the race wars of the American literary and political imaginary – situating, today, the rise of the "white supremacy" meme and its restoration of full Plantationcene protocols and decor, solicits a different hunt or tracking of the escaped slave within these impositions. A certain "Faulkner" offers a key to tracking petrolepathy into and through orders of blackness that refuse the mimetic and hermeneutic Plantation-habits whose normativity left it in place – hence the ease of the rollback today, in the era of climate autocracies and digital capture.

5 Trackings: Faulkner's Closure of *Ecriture* (*Go Down, Moses*)

> Faulkner writes in rhizomes.... To point out and to hide a secret or a bit of knowledge (that is, to postpone its discovery): this is a great part of Faulkner's project and the motif around which his writing is organized.... Is this some community we rhizomed into fragile connection to a place? Or a total we involved in the activity of the planet? Or an ideal we drawn in the swirls of a poetics? Who is this intervening they? They that is Other? or they the neighbors? or they whom I imagine when I try to speak? These *wes* and *theys* are an evolving.... [Faulkner] is not caught up in dreams of a racial panacea. Using everything at his command, he wants only to ground the enigmatic relation between Blacks and Whites in a kind of metaphysics.... the Faulknerian whirlwind.
>
> —Édouard Glissant, *Faulkner, Mississippi*

I

Tracking – pursuit of dispersed scents and marks, detective work – inhabits archival and critical labor, the gathering of textual clues and adjustments within reading along which something that memory has reason to obtain, secure, incorporate or unriddle. In Sophocles's satyr play, *The Trackers*, the priapitory ones, half-animal and half-man, reel about in slapstick, as if the formal step beyond "tragedy." Tracks fold back, collide, stop dead at rivers. Yet if memory were to track "tracking" itself, what seemed pre-originary like some prehistorial forest bear ("The Bear") or the referential regimes it is inscribed in, then the protocols of a "reading" might turn against themselves, opening the pre-recordings, momentarily or virtually, to disinscription. One of the many threads of J. Hillis Miller's work clearly circles this question. Endlessly. That is, that of a translation within, and of, not only programs of reading – which is to say, always also, perception – but of time, agency, memory. We might pretend to isolate two of Miller's positions – that of patient explicator of deconstructive methodology and that of an active intervention in the archive – and suggest that the latter, too, must be tracked moving behind the former. It is in his recent essay on Proust in *Black Holes*, coupled with an earlier piece on Faulkner and ideology, that one may get a glimpse of this at work. In the former, Miller deploys a "fractal" reading of Proust's diversely linked passages and introduces the book with a theorization of "The Fractal Mosaic."[29] The "fractal" is a convenient term for the multiple cuts and segments of writing that form the active cauldron of netherarchival space – down to the micrologic. This ever-recomposing "mosaic" is legislated, in its virtuality, by cuts of its own. Keeping the analog of literary telemnemonics and physics in play that the book's title promises, Miller invokes the prefigurative dilemma of "black holes" – the non-sites within the texture of a territorialized universe where stars have imploded, taking with them time, light and matter. He does so to position an impasse and turmoil within representation or tropological constellations, specific to an individual and perhaps micrologic archival site, which nonetheless takes on the properties of an agency of translation (much as a fabled "black hole" engorges itself with whatever comes into its proximity).

One might be tempted to insert this agency into Benjamin's figure of constellations of mnemonic ganglia and events, which would be seized upon at an utterly singular juncture (Benjamin's "monad"), making possible a rewriting of virtual pasts, as though to open alternative futures. The relevance of Benjamin's theorizing act mimes a dis-inscription. It is, today, perhaps a virtual performative above all. The intent is to alter and access the allomorphic nature of mnemotechnics, reference, as to intervene in the very contour of time, fore and back, politically and (above all) epistemologically. Since today's mnemotechnics is swarmed and pre-programmed digitally and increasingly through A.I. agencies and escarpments; and since the entirety of today's practices can be read from and against the inability to respond to the advance of *climate chaos* (which it – in the *Trumpocene* – occludes and banishes from the "real"), one may gauge the dumbing down of the (say, American) socius as impaled in a spellbound if encompassing site. There is no *pharmakon* for this herding of "consciousness." The "black hole" is the figure Benjamin could not wholly name, an agent of what sustains or affects the implosion of a spatiotemporal model or constellation. The "black hole" would not be an aporia passively defined, an indecidable figure, but the wormhole through which a certain translation and archival event is made virtual. Sometimes engineering a break in an epistemo-political program (or reading) is referenced to "ideology," as if that named, still, one manner in which ideation is logologically inscribed, then phenomenalized. In a discussion of ideology referenced to de Man and Althusser, Miller isolates for commentary an obscure passage about relays, translation and transvaluation – what is called "overpassing" in *Absalom, Absalom!*. Miller wants to explore in Faulkner where ideological investments operate within a machinery of discursive memory, as if they were inscribed pre-recordings that may be intervened in.[30]

The question is always that which accompanies the scandal of being preinscribed in a discourse, history, bloodline: how is a pre-recording, or pre-mapped reading and writing model, interrupted, opened to renegotiation or intervention by the very community that is programmed by such – dwelling in or produced by that archive to begin with? The latter suggests a movement or crossing that will link Miller's trajectory to one of Faulkner's own. Miller observes that in the relay networks of (a) reading, there is the possibility of an event, a decision, wherein the legacies under transmission do not just replicate old laws, rituals, ideologies or reference systems, but perhaps emerge otherwise: "What passes through that space, the space of decision, comes out different on the other side. For that difference the reader must be held liable" (208). "On the other side": a transvaluation, ethically cloaked. Miller is addressing *Absalom, Absalom!* and locates a passage that suggests one role of the term "was" in entering a network of traces, inscriptions and events:

> it would be something – a scrap of paper – something, anything, it not to mean anything in itself and them not even to read it or keep it, not even bother to throw it away or destroy it, at least it would be something just because it would have happened, be remembered even if only from passing from one hand to another, one mind to another, and it would be at least a scratch, something, something that might make a mark on something that was once for the reason that it can die someday, while the block of stone cant be is because it never can become was because it cant ever die or perish. (127-28)

Miller goes on to cite another text where such "a mark on something that was once" moves toward a transformation, or translation, which the reader must counter-sign. What virtual reader? Taking "ideology" to name where installed codes of reference (pseudo-realisms) are taken as natural, hence transmitted passively, he notes how the (Nietzschean) term "overpassing" – like the German *übersetzen* – conjures a hyperbolic signature of passage, along the lines we might call, with the fractal in mind, (a)mosaic. As a matter of translation, terms appear evacuated and reassigned roles beyond semantic pretext. Miller observes how the term "love" appears in the phrase "overpassing to love" in a cryptic way: "Love here is the name for a relation to history and to other people that may transform ideology and provide the glimpse of an escape from it" (214). *Love* appears an inverted or non-figure:

> it was not the talking alone that did it, performed and accomplished the overpassing, but some happy marriage of speaking and hearing wherein each before the demand, the requirement, forgave condoned and forgot the faulting of the other – faultings both in the creating of this shade whom they discussed (rather, existed in) and in the hearing and sifting and discarding the false and conserving what seemed true, or fit the preconceived – in order to overpass to love, where there might be paradox and inconsistency but nothing fault nor false. (316)

I bracket for the moment a certain logic pointing to Nietzsche – to a "going under" that is also a "passing over," to a too vague rhetoric of a transvaluation ("hearing and sifting") that might emerge here. This "overpassing," or *Übersetzung*, involves an alteration in and of words, of mnemonic traces and reference or readability: this occurs to the word "love," but also "faultings," and "false," and so on. What is called "the faulting of the other" would not be a casting of fault as guilt but a rifting, a networking of the radically other and nonhuman. It is glossed: "faultings both in the creating of this shade whom they discussed (rather, existed in) and in the hearing and sifting." A network of transfers, of passings-over as a seancing of fractal chronographs, "sifting," discarding: here the inscriptions that program a representational epoch stand to be suspended; here the "literary" emerges as a virtual epistemo-political "event," self-mocking yet proto-Mosaic before an arrested historical legacy and program. To speak of being in arrest includes, here, both a policial and hyperbolic inflection. "Faulting" here shifts from blame of the human other – a system or program of re-sent(i)ment, a repetitive reorganization of a past, of anteriority per se ("Was") – to a de-faulting of otherness that opens onto a non-anthropomorphic system passing through animal, "slave," thing, animation or life as linguistic or mnemonic effect. Thus, the word "faulting," in retaining Faulkner's signature effect (faul, fall, catafalque, going down), marks itself as a representational turn or backloop within a spool of temporal trace-chains. Does such a fractal or mosaic approach aid in addressing the politics of race or the animal, of blackness in "American" writing, or their purported transvaluation?

II

One may speak, then, of a certain zoographematics – that is, of how the categories of living and dead, animal and "human," species and effect, hunter and prey are organized and, perhaps, programmed through writing and reading systems or interpretive regimes. It is not (only) that tracking, camouflage, allomorphic metamorphoses and chemical defense participate in elaborate reading or signifying exchanges that seem to materialize alternative models of body, sensation, mutation, or even mimesis than our conceptual hegemonies can tolerate.

If the "animal" is routinely abjected by the demands of the "human" – positioned as its other, its pet, its subhuman, its terror, its food source, and so on – it seems sacrificed not only to a practical but to a semantic or archival drive. In this asymmetrical exchange, what is represented as the animal may be allied less to a code of the "natural" from which the human must be distanced than a techne itself, a phenanimality, let us say, an (a)material trace upon the occlusion of which a fable of human interiority depends.[31] To cite, as Faulkner will, the Negro spiritual "Go Down, Moses" as an enigmatic book title in order to interrogate and reperform the fall or disarticulation of this system – in association with the animal and race, and as a reading of "Faulkner's" production – raises the question of what resonances the name "Moses," or for that matter "going down" (like, in Miller's analysis, "overpassing"), may bear.

The alliance of a non-anthropormorphic, non-auratic otherness or black hole of tropological constellations with a kind of (a)materiality or inscription suggests other impasses. As long as beasts and runaway slaves are the subject of a hunt, a chase, of tracking or reading, only to be formally returned to the pen, the "animal" can be used as predictable excess, woven into hermeneutic rituals that parody closed programs of meaning. One tracks, one captures or not, one trees, and then one tracks again, repeating the same outcome, as if by a ritual of memory or concept, and again. If the ritual is rigged, chance all but defeated, the home protected again, one at least goes out again on the ritual hunt. Or almost, since this is also a juncture where intervention seems possible, or the "event" of translation forecast. Animal tropology remains, perhaps, a misnomer if what the slave or "animal" signifies may be also prefigural, premimetic or allo-anthropomorphic. The animal seems linked, too, to the proto-Mosaic prospect of an alternative reading model – which is to say an alternative epistemo-aesthetic regime to the tropology of the "human." This subjects the figure of the "animal" to a machinal inscription, turning up within the catalogues of the hunt, the chase, of tracking – even where, as is obvious, the cluster of animals more or less close to "man" in this regard, and variants within a form (say, dog and fox) occupy both sides of the chase, and help turn that into a game without decisive division of hunter and hunted: both are made, in the service of "man," to obfuscate the circulation of mastery. It is enough, apparently, to defer the question indefinitely as the human regime or semantics proceeds. This animal's prefigural role as at once trace and bearer of sheer anteriority contaminates, in Faulkner, the position and logic of *the runaway black or slave*.

Just this *was*ness – the order of anterior inscriptions on which alternate pasts and virtual futures, alternate reading programs depend – seems gambled in the labyrinthine, cross-wired texts spanning a century of Faulkner's *Go Down, Moses*.[32] The title itself declares a fall of and within an entire history of symptoms it offers itself as exposure of. It declares the semantic or Mosaic law which the authorship, "Faulkner," emerged within as closed, what is reflected in the arrested plantation economy whose impasses and sterility frame the opening tale, bluntly named "Was." The word "was" (originally, it was called "Almost"), much less evokes an antebellum past than perhaps furiously conjures sheer anteriority or posits as past or over an entire representational system. The paralyzed plantation of the twin brothers stands between an absent trauma and retraction of all paternity ("Old Carothers") and a (civil) war, a historial intervention and rewiring, good or ill, of the archive, of property, of legibility – hence, the oscillation between achieving a near event (almost) and representing that as perpetually anterior ("was"), a repetition. It marks this arrest, this arrest of arrestation or blind pause, through the image of the ritual hunt, of tracking – used, inescapably, to imbricate a hermeneutic model of reading run aground in repetition, formalism, its own management of anteriority (that the animal, as trace-figure, evokes sheer "anteriority" also remains a lesser puzzle here). The text wants to sacrifice itself, as the character Isaac later does all of his property, toward some other proto-Mosaic crossing it cannot name. It is caught in this inter-space, between a foreclosure and a pointed-to beginning, between "Was" and "Almost," and in this, it turns to animals, and, of course, blacks. The book's title cites not only the Negro spiritual exhorting the liberation of the slave, of blacks in the social history of the South, and so on. It shivers, splinters against itself. It stutters, like a lisping, double Moses, unable to affect this crossing ("Was" turns back altogether before the event, the card game, the impending "civil war," and so on). Here Moses twice goes down or under (in the Nietzschean sense), who brings laws and casts laws down (shatters them), then comes down again with their simulacrum, an authorial Moses who can only indicate a crossing to another historical system he seems barred from.[33] Three questions guide this pursuit – or, rather, a reading of the structure of "reading" that is shaped by the narrative of the hunt, of the hermeneutic chase, of the "animal":

1 How do the sterile hermeneutic rituals of the "twins" reflect not an antebellum idyll, but repetitive and formal

hermeneutic rituals which we, not far from the mid-twentieth-century, wartime Faulkner, still inhabit?

2   Why is the only mention of the name "Moses" in this writing attached to a vaudeville or slapstick hunting dog, who comes up with his head in an empty crate?

3   What is implied by "treeing" in this work – that is capturing the lower order of the trace, of the runaway slave or animal, in the heights of a tree, with no exit, in the place of leaves, of natural emblems or referents (Saussure uses the image of the tree to illustrate the divide between so-called signifiers and signifieds)?

Go Down, Moses – whose tales refuse to integrate into temporal narratives, family histories, or even stable genealogies – can be said to identify in and with its opening performance, "Was," an arrest or even relapse in time, an arrest of the arrest of time. One occurring not in some fantastic history of "the South," of antebellum plantation life and frontier comedies, so much as within the Mosaic or hermeneutic labyrinth and a stalled set of reading or mnemonic protocols, and for that matter a stalled model of the "aesthetic" itself, which puts literature, history, the future and itself into question. It is symptomatic that the once canonical reaction to this late work routinely considered it a relapse of sorts, a return to nineteenth-century realisms, a book of personal mourning and guilt, an homage to black folk underpinning the "South," and, at the same time, a work of decline.[34] But a double arrestation marks the plantation model of Uncle Buck (Theophilus) and Uncle Buddy (Amodeus) – *twins*, each with two names, marking not so much dialectical pairs as giving the lie to such (as if to say, troping the pillars of literary history, the dyads of Greek and Hebraic, aesthetic and theologic, symbol and allegory were, all along, the same), sterile old men living like husband and wife, maintaining the fiction of slaveowning they also deny. The twins' ritual of slaveownership, which they perpetuate while disavowing (a typical critical imbrication), depends on a fiction that permits denial. Like Uncle Buck's dressing up to court Sophonsiba while pursuing the runaway Turl, openly rejecting the very prospect, we later learn that the twins not only moved the slaves to the (unfinished) big house but, while latching the front door leave the back open at night, for the slaves to go where they want as long as they return by dawn – pretending, equally, not to have left. "Ritual," as with the foxhunt inside the house, which recurs without consequence, is wedded here to an arrested set of interlocking fictions, aestheticized routines, which preserves a long, derealized state. Less the antebellum South than an epistemo-political impasse of a (still) received model of "history," Uncle Buck and Uncle Buddy – Fauna and Flora – are brought to the very brink of a translation ("Almost"), even as it is already past as well ("Was"). In the story "Was," and throughout *Go Down, Moses*, the text tracks (itself as) a shift from one system of writing-reading – one set of stalled, aged, disempowered inscriptions – to another, by exposing the techniques of that "Mosaic" system (the rituals of property, doubles, homosocial power, slaves, animal determinacy, impasse) to disinscription, to an "outside" represented by chance, and hence to an alternate future: only it is not the "antebellum" South at issue but the (still perpetual) "present" of a mimetic regime or culture.

III

The work performs its narrative of escape and return as a doomed hermeneutic ritual even as "Was" asserts that the genre caricatured – not just historical fiction, but the "literature" it caricatures – is past, over, a dead commodity. The narrative opens with a ritual hunt, only here the dogs are loosed on the twin's pet fox inside the house: a repetitive mnemonic game without consequence, over almost as soon as it starts, enjoyable ("It was a good race," we hear). After introducing the disjunct relay system of voices by which an anecdote of the antebellum past ("out of the old time, the old days") was passed to Isaac McCaslin – that of his to-be father, Uncle Buck (Theophilus McCaslin), and the latter's twin brother, Uncle Buddy (Amodeus McCaslin) – "Was" turns to a repetitive, staged "race" occurring within the bizarre household, or economy, of the sterile male twins living almost as man and wife. Over all of this, of course, the question of an "event" either gathering (the *Civil War*) or just disturbing this time-frozen site haunts this historical and cultural program or (a)topos, this doubly fictional plantation (the fiction of slavery, for instance, will be maintained even as the slaves are let to roam at night, returning before dawn as if by agreement for appearance's sake, even moved into the unfinished masters' house, the brother-masters at once abdicating and maintaining these rituals). The scene prefigures and awaits the catastrophe of war, or is in residue from the disaster of the absent patriarch, dead, "old Carothers," in whose residual accomplishments and crimes the twins more or less dwell. All "dialectic" is at a seeming impasse in the image of the old twins. An opening "race" is featured – dog chasing pet fox, predictably and as sport, in the house itself, with a return to the original position, again and again, as play – that further collapses any interior of the house. It then gives way to yet another ritual pursuit and retrieval, this time that of a runaway slave, Tomey's Turl, half-white, and half-brother to the twins. Based on this ritual pursuit, the possibility of Uncle Buck's reluctantly courting and being caught by his female neighbor, Sophonsiba Beauchamp, is raised and, within the tale, barely evaded again. It is another key event, if it happens, since the entire book-to-come and non-novel will depend on that character, that relay-voice, that son and not-son, that memory trajectory and his putative and ineffective sacrifice – later, of property, out of guilt of sorts, or responsibility, or rupture. The word "again" echoes a maimed, eternal recurrence, "old" software caught in a loop:

> When he [McCaslin, the boy's elder cousin from whom the tale is relayed] ran back to the house from discovering that Tomey's Turl [the half-brother slave] had run again, they heard Uncle Buddy cursing and bellowing in the kitchen, then the fox and the dogs came out of the kitchen and cross the hall into the dogs' room and they heard them run through the dogs' room into his and Uncle Buck's room then they saw them cross the hall again into Uncle Buddy's room and heard them run through Uncle Buddy's room into the kitchen again and this time it sounded like the whole chimney had come down and Uncle Buddy bellowing like a steamboat blowing and this time the fox and the dogs and five or six sticks of firewood all came out of the kitchen together with Uncle Buddy in the middle of them hitting at everything in sight with another stick. It was a *good race*. When he and Uncle Buck ran into their room to get Uncle Buck's necktie, the fox had *treed behind the clock* on the mantel. (4-5, my italics)

I ignore, for now, the marking of sheer or trapped recurrence at the start ("again ... again"), or the treeing that occurs, we are told, "behind the clock on the mantel." Treeing. That is to say, being trapped in the upper part of a tree where no escape is possible, a ritual here indicating crudely a suspension of time (did the "clock" not itself rest on the "mantel," gesture toward a telemnemonics and historial switchboard, like the narration, for which the twins, or man, is cipher). One would be tempted to draw a connection between the animal hunt and the hermeneutic ritual of reading within a programmed model, the slave figure in this case the (a)material carrier of sense, prefigural markings, inscriptions, the "was" as such.

The text engages its narrative of escape and return as doomed ritual, much as "Was" asserts that the genre caricatured – not just historical fiction, but mimetic "literature" tout court – is past, over, a dead commodity. The system, at a point of impasse, will appear brought as if to an encounter with chance: that is, a place where its apparent formalization will be displaced by engaging a factor as if outside, a point of intervention. This will appear to occur – if only, again, to retreat from the "event" – in a poker game in which not only Tomey's Turl but Uncle Buck himself is wagered (the latter, caught by Sophonsiba Beauchamp).[35] On this non-encounter, mimicked by the ritual chase and retrieval of the fox, various futures hinge. Repeatedly, and across the tales of the "book," there are encounters with "chance" (often in [fixed] games), which suggest the formalization of a ritual system to the point it is at risk of exceeding itself, or contact with an external (chance) factor, an archival intervention (in this sense, "Was" names archive).

In "Was," as elsewhere, this circulates around the figure of the "hand," both as the dealing hand and the hand of cards dealt – a figure of agency, as in "The Bear," that not only is crossed with the writing hand (or "His"), but cites the first prehistoric aesthetic markers in cave paintings, the digitalization that at once signified the (faux) human and technicity as such. In "Was," for instance: "'You took your chance too,' Uncle Buck said…. 'One hand,' he said…. 'The lowest hand wins Sibbey and buys the niggers'" (22-23); "and the hand dealt him an ace and Uncle Buddy a five and now Mr Hubert just sat still… and watched Uncle Buddy put one hand onto the table for the first time" (26), and so on. The coupling of Buck and Sophonsiba – on which the future narrative depends – lies in the balance: and does not occur. (Uncle Buck's name cites both the male deer who elsewhere appears in the work as an ancestral revenant and the book.) The future which is prepared here, and which the novel testifies to, also does not occur: the entire narrative of Isaac, the impossible non-sacrifice which mirrors that as if by one Faulkner-Moses of his own authorship. In fact, one could say that the coupling does not occur – that the text sacrifices Isaac before his non-birth, a ghost in advance, unable, in turn, to sacrifice himself through the sort of auto-dispossession of legacies on which gesture "The Bear" ineffectively turns as an iconic attempt to undo the historical chain of debts, "history," indeed, what is called "the Book," or what the work repeatedly calls "trace-chains."[36]

For economy's sake, one may be less interested in how the homosocial white male order is presented as a sterile, perpetual fiction of semaphoric property than in where the "going under" of an entire archival regime appears at risk. In this way, works make contracts with virtual, non-existent readings, targeted at immaterial futures. The chase, of course, ritually returns what is hunted back to its place – the fox, the slave, Uncle Buck – even if with a twist, a remainder. We recall that Plato uses the trope of the chase or hunt as an originary trope for the pursuit of truth, of meaning, and hence reading itself. The closing chase of "Was" echoes the opening slapstick dog hunt in the house – the trope of all mere "literature," of all writing that pretends to discover the revelations it planted:

> And when they got home just after daylight, this time Uncle Buddy [that is, Amodeus] never even had time to get breakfast started and the fox never even got out of the crate, because the dogs were right there in the room. Old Moses went right into the crate with the fox, so that both of them went right on through the back end of it. That is, the fox went through, because when Uncle Buddy opened the door to come in, *old Moses was still wearing most of the crate around his neck until Uncle Buddy kicked it off of him*.... And they could hear the fox's claws when he went scrabbling up the lean-pole, onto the roof – a *fine race while it lasted, but the tree was too quick.* (28, my italics)

Why is the cartoon-like hunting dog, here, alone in the whole book given the mock-authoritative name "Moses" and *old Moses* at that, as if designating a ritually comic model of hermeneutic pursuit, mock-authority, animal repetition, anteriority? Why, once again, is the unnamed fox that is the quarry – parallel to Tomey's Turl, the half-brother slave ritually chased and returned to the plantation, though in fact guiding the whole operation and the only chance of the sterile twins for a future at all – treed? And why treed in a fashion not only called "too quick" (a degree of pleasure in repetition now lost, too easily done even for that, the predictable closing in on itself), but first praised (again) as a harmless "race," an aestheticized hermeneutic ritual. Alter the hermeneutic program of mnemonic pursuit and return to the plantation, the "race" that is an effect of this double chase as well, this ritual regime, and one disperses the referential program and archival drive to enslave the carrier, the bearer (or bear). In the territorializations of race, of the "black" whose renderings are said to return the book's performance to a retro-realist narrative – rather than a hyperbolic intervention within the machinery that produces those reading models – the text locates the agency of a black hole, a blackness precedent to figuration that is neither white nor black precisely.

IV

"Treeing" technically involves a chase, a hunt, a pursuit and retrieval: the pursued is treed, driven to the highest branches but there as if trapped. The lower or slave figure becomes the higher or signified crown. Hyperbolically trapped. One is caught in the upper strata or branches, like the squirrels trapped at the close of "The Bear," running about like "mad leaves," while the inept hunter Boon sits below, hammering at his "dismembered" rifle, metal against metal, calling out "Dont touch a one of them! They're mine!" (315) (trying, still, to impose a system of property, of meaning or *Meinung*, upon this translational event):

> At first glance the tree seemed to be alive with frantic squirrels. There appeared to be forty or fifty of them leaping and darting from branch to branch until the whole tree had become one green maelstrom of mad leaves, while from time to time, singly or in twos and threes, squirrels would dart down the tree then whirl without stopping and rush back up again as though sucked violently back by *the vacuum of their fellows' frenzied vortex.*
>
> [Boon] didn't even look up to see who it was. Still hammering [on his "dismembered gun"], he merely shouted back at the boy:
>
> "Get out of here! Dont touch them! Dont touch a one of them! They're mine!" (315, my italics)

This limit moment of treeing, framed by Boon's imaginary propriety, is allied to the most minimal, or even preoriginary, repetitions of sound, mechanical, "dismembered," metal on metal, what might almost evoke (one cannot say "be represented by") a prefigural seriality of slashes or bars, of tracks, what allows the word "treeing" itself to echo further – (t)ree-ing, t(h)reeing, (t)re(ad)ing. If the prefigural is that out of which phenomenality is projected, the senses programmed, meaning installed, only a rupture or arrest before what is preoriginary opens virtual dis-inscription.

For example, the word "race." This Mosaic network at the point of going down or under trades, here, on the word "race" almost unacceptable – for the term, woven into the genealogies of white and black here, appears only in this aporetic form, this ritualization and caricature of a (de)faulted reading model or chase: as though the entire representational model of "race" (racial difference, slavery, animality) were itself generated from this hermeneutic ritual chase or tracking, from a compulsion to find a designated prey, and not the other way around. We must hear the injunction to "go down" in the perhaps Nietzschean sense of "going under," according to which an entire representational and experiential system stands to be dissolved, gambled, put in dangerous contact with its exterior. And that, finally, at a point where its sheer repetitiveness, its utter predictability makes of it a merely aesthetic pleasure ("a fine race ... but the tree was too quick") – as, in "The Fire and the Hearth," the treasure-finding machine of Lucas "discovers" gold coins that had been pre-buried, planted mnemonically for the theatrical effect of discovery or recognition, like the regional pleasure taken in a certain "Faulkner's" historical *fiction writing*, and so on.

In a text of twins with double names and double "Warwicks," we can now distinguish between two virtual uses of the term "aesthetic – "the one, that of a faux "literature," or "old Moses," accords with the mere aestheticism of a ritual of memory formalized dangerously, upon which the plantation semantics of the male twins' sterile and arrested mastery (and notion of the human) resides, but the other is allied to the treed fox, the animal as trace, prey or agent.[37] This more preoriginary problematic of the "aesthetic" is linked, we might say, to the site where perception (*aisthanumai*, in Greek) materializes together with an effect of reading linked to what may be called an (a)material trace – like the movement of the subhuman slave, the fox accords with the lower order of a binary model, like the signifier (or inscription) effaced before the pretense of the signified (meaning, the upper portion of the tree). To be "treed" performs a limit movement in the going under of the old Mosaic order – of the twins, the human, the animal, the ritual hunt, white semantics, "history," and so on – preparatory to a crossing, or crossing over (*Übersetzung*), proto-Mosaic yet already performed by the text, which, among other things, traces the passage from a model of trope to a prefigural and performative model in which animals, time, race, justice, sacrifice and the archive itself are at risk or, again, gambled. Gambled how, and toward what? And if as a poker game, played with hands, with what recourse to bluffing?

It is here that the imbrication of animemes – bee, buck, hawk, dog, fox, horse, buffalo and, of course, bear – in the order of the slave, the nonhuman carrier or *bearer*, emerges, much as the phoneme "bear" implicates the Greek *-phorein* or *-fer*. Yet this raising of an (a)material order of the signifier to the upper order of the "tree," trapped there, renders the letter the signified, and short-circuits (almost) a set of protocols that sustain the Mosaic model of property, the natural, the semantic, metaphor, "literature," and so on. It is in the work's last tale, from which the book's title is taken, that this accord with a problematic of arrested translation is most clear – *translation* in a Benjaminian sense, where a sort of "pure language" (*reine Sprache*) shimmers which inversely denotes a sheer *materiality* of signifying networks precedent to any meaning effects. Such a *dispossession of plantation semantics* accords with a potential reconfiguration of the historial, of its virtuality, precedent to an event ("Almost") or disinscription ("Was"). Any interrogation of the role of "Egypt" in "Go Down, Moses" would necessarily pass through the problem of "the Book" (Hebraic) and (Greek) letters. But the regression of the Greek *Delta* to the inverted pyramid or triangle breaks out of the Anglo-alphabet less as Greek letter than a geographic design on a page of "Delta Autumn," and Aunt Mollie's bizarrely reversed chant about Pharaoh and Benjamin in "Go Down, Moses" ("Sold him in Egypt and now he dead," 363). In a certain meltdown of historial periods the text(s) perform, "Egypt" names a site or writing seemingly without precedent or successor, the cover for a preoriginary or graphematic "constant" upon which other epochal narratives are cast and absorbed, projected and disappeared.[38]

V

This *law* and this *translation* are noted. Of Gavin Stevens it is observed that the law was his "hobby" and his "serious vocation was a twenty-two year old unfinished translation of the Old Testament back into classic Greek" – that is, a translation of the ur-Book's (English) translation as if "back" not into an originary script but into the aesthetic language which, nonetheless, precedes Mosaic law as trace: a virtual backloop, scriptive and temporal, to a language of performance from which the world and history are as if materialized, phenomenalized. The prefigural domain of *(dis)inscription*, black holes. Here the "liberation" of the racial other within the story of American slavery and of the animal other, within the protocols of homosocial "human" definition, is directly crossed not only with a shift in the contract of reading-writing as such – within the agency of material language and its mnemotechnics – but the order of time. It is in another text of this deadly volume, "Pantaloon in Black," that the mimetic regime is cast off by the titanic black laborer Rider in what comes closest to a performative meditation on such a "crossing" (as it is called). The name "Rider" unexpectedly mimes, incorporates and fuses the agencies of "reader" and "writer" both, and what he traverses in a scene beyond mourning, as we may call it, is not only where the vigilant "white nightwatchman," *Birdsong* (a certain seductive pretense of metaphoric literature) fixes the *game of chance* for the black laborers, but a certain technicity precedent to this site ("the door of the tool-room"). The name "Birdsong," in fact, occupies a double fold: on the one hand, it cites the cheating nature of traditional aesthetics as a seductive site of mnemonic manipulation (like the treasure-machine), and mystification, which Rider crosses as if from; on the other hand, as in the opening of *Sanctuary*, where a "carolina wren's" birdsong is allied to the repetition of three notes alone, it is a wordless form of (a)rhythm, a bar-series, a marking system precedent to metaphor, sheer technicity:

> He crossed the clearing and entered the boiler shed and went on through it, crossing the *junctureless backloop of time's trepan*, to the door of the tool-room, the faint glow of the lantern beyond the plank-joints, the surge and fall of living shadow, the mutter of voices, the mute click and scutter of the dice, his hand loud on the barred door. (147, my italics)

I ignore the violent contortions which the *atopos* word-name "trepan" necessitates (an active black hole as a letteral assemblage). Yoking "tr(h)e(e)" in all its resonances with the totalizations adhering to "pan" (or "panta"), it also suggests an intervention, an operation or surgical cut at the heart of memory or identity, supposedly, in the skull and brain itself. This totalization is marked in "Pantaloon in Black"'s opening sentence by the word *"overalls"* ("He stood in the worn, faded clean overalls which Mannie herself had washed"), a hyperbolic inflection in which the "over" countersigns the "going down (or under)" trajectory which *Rider*, more than any other character in the "book," bears, a trajectory that begins, at its opening, as though "beyond" mourning, or thinking understood as the relapse solicited by a mimetic model that confirms and reinscribes mourning as the defensive mechanism of "human" representation. *Rider* will toss a giant log(os) in the air, as if gravity-less, at the saw-mill – a cutting of trees, a production of pulp paper, a linkage of sight (saw) to this cutting and to the inversion of the opening tale title ("was").

"Crossing the *junctureless backloop* of time's tre(e)pan" registers a precession, perhaps, of the entire programming of perceptual order, of one archival and temporal template, of the human as programmatically defined. It is why the animal affiliated Rider ("they aint human," says the Deputy: "They look like a man and they walk on their hind legs like a man, and they can talk, ...(but) they might just as well be a damn herd of wild buffaloes") unpacks the unexplained title word, "pantaloon," in unreadable ways: not just as a commedia dell'arte figure ("pantaloon"), in which a certain pantomime, or panta-mimesis, echoes, but whereby the "pan-" or "panta-" itself bleeds and not only reconfigures again in a "panting," or breathing (*spiritus*) that erodes the difference of inside and out or interiority, but delivers us to a phantasmatics in which phenomenalization (*phanesthai*) intertwines not only with the ghost but the *pensere* of thought or thinking itself. Rider's last, prelynching utterance discloses his entire trajectory, unexpectedly, as an errance of thinking: "Hit look lack Ah just cant quit thinking" (154). *Thinking.* That, for Rider, the recurrent "Ah" replaces the first person inscribes this subject-performative as a perpetual prosopopeia of that position itself – a formalized precession of the "human" or "Mosaic" or authorial precept. "Thinking" – a mutation of inking and in-in(g) – names the directionless direction of Rider's post-mourning errance, of the text's own crossing, as a mutation of the very model of an (interminable) thought or form of thought. It is here that "treeing," in the sense of lynching, disarticulates not only race – except to rewrite "blackness" as a pre-photological topos – but animality, and the aesthetic as the site of a phenomenalization upon which a crossing of epochal signifying orders (and "Man(nie)") stands to be gambled. The white twin brothers, Amodeus and Theophilus, it should be noted, by no means constitute a "patriarchy" but, rather, something of a perpetual semantic eunarchy that is routinely rewritten from after the fact in patriarchal logics – it represents, if that word can be used, with its emphasis not only on the homosocial fraternity but on sterility, eternal recurrence and arrestation, a vacancy that the allusions to "old Carothers," the always already absent criminal-incestual "father," nowhere fill out.

This *crossing* unnames the performative turn against itself of the entire "Faulknerian" writing machine and the "Plantationcene" default legacy – functional not as an antebellum Plantation åomedy, but as the spell and paralysis of a plantation mode of reading and hermeneutic crackers.

VI

The *going down* or (de)faulting of the Mosaic reading or hunting model, the white twins, the surgical "backloop of time's trepan," the faux trope of "the animal" to cover a terror within inscription itself, occurs where the so-called aesthetic stands to be transformed according to a prephenomenal model of inscription, mnemonics: "treeing," here, becomes a homonym for the *techne* of translation according to Benjamin, which resembles "treeing." This entails what he calls turning "the symbolizing into the symbolized,"[39] a shift in which the carriers (slaves, beasts) of signification are turned upon, taken up, replacing the imaginary signified with the technicity and (a)materiality of signification. If what is called "allegory," or so Benjamin notes, negates that which it represents, the Faulknerian text negates or disowns its own production as a means of "overpassing." "Tree" summons (and dissolves) the genealogical model with its branches and roots, even as it summons (and dissolves) the problem of the *natural* image, that is, the referent of mimeticism that depends on figures of the earth, of world. One form this translational movement (up) takes, the treeing preparatory to any translation or crossing, is a shift from an imaginary referential code monumentalized as "history" – the nineteenth-century naturalism Sundquist supposes, say, the point of the writing's relapse – to the performative event of ((a)material) words and trace-chains, words which transfigure the whole text in demanding to be "read" as prior to phenomenalized thinking: trace, heart, chance, spot, tree, bear, and so on. It performs in advance an unreading of the Plantationcene as a hermeneutically closed and entropically deteriorating static economy, a "reading-writing" program in default. When "Was" describes the two separate "plantations" of Hubert and Sophonsiba Beauchamp – Miss Sophonsiba's is named "Warwick" in a Walter Scott-like mode or fantasy – we hear: "it would sound as if she and Mr Hubert owned two separate plantations covering the same area of ground, one on top of the other" (9). That is, a split between a literary or mock-allegorical signifying order and another real, of which the first accorded with the "literary" construction of so-called historical ritual. Faulkner has gotten entangled in the labyrinth of a "plantationcene" at once horizonless and in utter default.

Viewing the iconic but elusive tale at the center of *Go Down, Moses*, "The Bear," as itself a reading or furtherance of "Was" produces a somewhat different text than that assuming a mimetic *Bildungsroman* (what does it mean, anyway, for (an) "Isaac" who will try to dispossess himself of his own (literary) legacy to be tutored by an ante-historical animal figure, carrier or bearer of sense?). For instance, that the anticlimactic killing of "Old Ben" (a redundant name for sheer anteriority, old be(e)n) strangely mimes the *Laocöon*, icon of the violence inhabiting the Greek aesthetic – site, that is, where what has been bracketed as the "aesthetic" ("a fine *race* while it lasted") is rewritten as the bearing movement, let us say, of a material and prephenomenal order. For it is not by contrasting a fictitious human subject to a projected animal subjectivity that one accesses, say, the trans-species "other." Rather, this is done by divesting a received regime of the "human" which may never have existed by what upholds its plantation – how, say, the "aesthetic" is positioned. Thus, yet another Faulkner, the second "Moses" that undoes the parody of the first one, with his "literature." (The age of "literature" as that of the animal's extinction.) Interestingly, the "earth" that is appealed to across Faulkner's mosaic text is not natural, not idyllic, not that of mere loss or mourning or recovery: "out of the earth, beyond the earth yet of the earth because [Isaac's] too was of the earth's long chronicle" ("The Bear" 297).

It is an earth *and* an ear, we may say, of the black hole or linguistic techne which precedes the lumbering chronicle of what is called "the Book" with a sort of pre-historial trace or "natural history" effect. *Aterra*....

Faulkner's zoographematics as one site of going under, or down, is referred back, as it were, to the (a)maternal figure to whom the work is dedicated. Here, in what exceeds any mourning for the maternal or even earth, the word-name "bear" is prefigured by the name "Caroline Barr," the black-mother substitute for the trope of natural (or white) maternity, the prosthetic "mother" as bar(r) or mark upon which "trace-chains" may be re-cast – a *khora* figure in the sense elaborated by Derrida of a non-existent or *atopos* site of disinscription and reinscription of programs, a "mat(t)er" that is herself neither woman nor man, the supplementary techne precedent to trope itself, a simple bar(r), at and of the origin according to the logics of which "life" itself is disclosed a telemnemonic effect. "In" black. In Faulkner's dedication, *khora* is called "Mammy," (echoed in the dead "Mannie" of Pantaloon, where, however, under the sign of a female corpse, "man" and "not-man" intertwine).

Mammy? The Black *techno*-mother? One could contrast and compare the phrase "going to earth" in "Was" to the auto-eruptive description of the burial mound – which is also that of the buried treasure in "The Fire and the Hearth" – opening "Pantaloon": "the mound seemed to be rising of its own volition ... thrusting upward out of the earth itself " (131). As the titular word "hearth" marks, "earth" participates in a trace-chain not of mystic or maternal origination, but an eternal recurrence of signifiers and markers, living and "dead," that intersects with ear and hear, art and heart, and so on, *trace-chains* "Pantaloon" perhaps notes as "insignificant to sight but actually of a profound meaning and fatal to touch, which no white man could have read" (132). Much, if not all of this, it seems, adheres to Faulkner's naming his comic hunting dog, caught in a hopelessly formalized ritual whose aestheticization is no longer efficient, is too accelerated for effect ("the tree was too quick"), and which marks by default a reversal of the term "aesthetic" itself as though pre-inscribed in the entire system all along – one upon which, moreover, the very definition of the "earth," of "going to *ear*th," hinges in this buckling permutation (or junctureless backloop) of and within and "old" writing and reading, which is to say chronographic, machine. All of this, in the most unobtrusive corner of a neglible story, a frontier comedy, perhaps (as the book's title suggests) about a hunting dog named "old Moses."

VII

On the *other hand* – a proliferation of dismembered hands occurs at this point – the skit also extends the text's performance, virtually, as a site of this going under, translation, "passage" or disinscription, of the sort a fully formalized system – at the point of arrestation or impasse – prepares "itself" for and in the process declares already, by the fact, accomplished, past-tensed, "was." If the era of "literature" as an institution – going under with "Moses," here, together with historical fiction (or the fiction of historicizing writing) – is that, too, of the extinction of the "animal," the era of the "human" as sterile or twinned white brotherhood and the hunt, the "end(s) of literature" which "Was" prepares or performs, say, suggests a shift, or passage, to another mnemonic network and model of the (mnemonic) event. Today, it may also imply that what was called "literature" institutionally may be rethought along the model of such virtual or tele-mnemonic or (an) archival events, dis-inscriptions and crossings, as though from a "tropological" to an actantial but irreversible occasion.[40] *Go Down, Moses* marks and, treed, lynched like Rider, disappears into one limit or preparatory reshuffling of this same sacrificial movement; caught in the impasse of ritual repetitions, of a seeming inability for the slave figure, or term, to run away, it also posits a faulted crossing to an other anarchival order or law (Moses, going down, gives [up] the [new] law).

What Miller's intervention prepares and alternately performs – while pursuing local pedagogic aims – includes this event and translation, bringing matters to the point they are, say, in the gaming scene of "Was" ("state of emergency," Benjamin tropes): where the outcome of a poker game may (or may not) set the terms for what is to come, what may (or may not) program an(other) epoch, or what was designated as passed, over, fact, history, "was." At issue in Miller, in Faulkner, can be termed an (an)archival *crossing*, a dis-inscription and deferred reinscription, the building of an allographical archive-to-come, which is at once virtual and already fact, which would recast an archival regime by which temporality and history, perhaps, have been legislated – hence, marking or indicating an other archive, a double or alloarchive that is not regionalist, not (only) "America," not regionalist or "humanist" in the model of the twin white brothers, the homosocial fiction of the white master – old Carothers, absent, Sutpen-like, catastrophic. Rather, dis-inscription is optioned when an archival program turns back on itself, as a performativity without circumscription, and comes closest to passing over – to posing the question of its own reading as a proactive transition into something beyond itself, which like a Moses (there are, clearly, a flock of different, competing, even multiple warring *Moseses* here), since at issue is another law perhaps, it can point to and stutter, effectuating what it cannot recognize itself as.

Tracking Miller's "black hole" through so-called ideology in Faulkner opens a curious vista, and a seditious one. Preoriginary to white or black, the site of an historical regime and aporia, the black hole of prefiguration locates itself in, and behind, different micrologic nodes – yet operates not as an indecidable marker or aporetic aestheticization but rather as an agency for re-networking the telemnemonics of scriptive legacies, the missing agency for intervention in Benjamin's archival constellations. The "literary" or the "book" becomes fractal, active and mobile mosaic chips on an allochronic switchboard of virtual events of reinscription. Translated fore and back in a junctureless juncture and proactive backloop, race and animal are transposed into other terms, those of a prosthetic earth. In the mode of the fractal theorized around Proust by Hillis Miller in *Black Holes*, however, one could say it is a mosaic of disjunct pieces crisscrossing temporal spaces, with the exception that these elements, pieces or historial monads can be proactive, resituating themselves in a mobile and erasing set of mnemonic constellations: "America," blacks, animals, historiality, the event and the Americanist regionalist bias that reinforces the plantation readings of "Faulkner." Returns the escaped half-brother slave, again and again, until "he" becomes the central actant, and beyond himself as unreadable agency wired to another blackness networked to the other side of the mimetic looking glass – networked to, entrapped within and distinct from the hermeneutic plantation rituals, sacrificing general histories (back through Rome, Greece, Egypt....) and "literary" authorship.

VIII

I quoted Glissant's *Faulkner, Mississippi* above, in which the peripatetic critic sculpts a remarkable narrative of a visit to place that interrogates Faulkner at a remove from identitarian receptors (he is concerned young blacks give up on reading Faulkner in the '90s). He re-enters the text from quasi-outside (Caribbean, Black, Francophone), and takes the mantle of travel guide literally and as reading premise. Grafting off the Deleuzian vibe of the day, he tells us to track *rhizomes*, sidestep *regionalist* Americanist interpretation, and return "Faulkner" to reading otherwise. He seances the outline of a broader, immersive "Plantationcene" – amplified as the go-to model by MAGA and the consolidating Trumpocene of climate autocracies and techno-oligarchies. Yet this also lies at the core of the *hermeneutic paralysis*, antebellum, that the aged twins Amodeus and Theophilus are caught in at once living in and abdicating their "plantation," a paralysis opening *Go Down, Moses*.... The *race* is ritualized, reading caught in entropic recurrence: a hyper-materiality that the writing of *Go Down, Moses* wants to deface, leave, escape, cancel through Black agencies, an escape which cannot be done by Isaac's own self-dispossession (mimicking Faulkner's), stamped as failed in "Delta Autumn" – learning new ways to hunt (read), or not quite ever.

Glissant's *Faulkner, Mississippi* convincingly conjures Faulkner as a "rhizomatic" writing – yet pursues a multiplaned tour of place, writing, a "Black and White" metaphysics, as if entering the "American" bubble from without, the perspective of a Caribbean, a Francophone, a Black. Yet if I piggyback on Glissant's strategy, it is with a different aim, much as rhizomes must be considered as scriptive.

Two emendations for the era of climate collapse, and where the "race" staged to mimic reading's hunt itself has expanded to that between two tempophagic vortices pitted against and feeding one another – AI and climate extinctions. What is overlooked is that *rhizomes* and the rhizomatic are mere tropes if not realized in progenerative networks, embodied, hyper-material, "semiotic," inorganic, *mnemotechnic*. Blacks within *Go Down, Moses* tend to cover a site of technics and scriptive agency having to do with reading or writing or telepathy or spectrography. They guide any *rhizomatics*.

It is Luster who drives the buggy against Benjy's anticipated left-to-right reading direction at the end of *Sound and the Fury*. (He says he is not going to the same "boneyard" as his mates, whupping his horse: "Hum up, *elefump*" (teasing the *Aleph* and opening a list of word-things outside of grammar ("..., post, ..."), which Glissant cites). Riding, that is, Faulkner's inversion of Plato's chariot in *Phaedrus*. Yet it is also a certain "black" stuff that Popeye (nominally white in *Sanctuary*) associated with his eyes and, for the lawyer Horace, with the stuff flowing from Bovary's mouth as corpse in the book he is reading ("do you read books?"). In *Go Down, Moses*, not only does half-brother slave Tomey's Turl alone guide the attempted rupture of antebellum hermeneutic paralysis, and fix the final poker game with Uncle Hubert, or open the prospect of a new generation of McCaslins, or at least the narrative voice of "Isaac" – whose failure to rupture family history despite all manner of renunciation, provides an arc to the volume's segments. Aunt Mollie *cannot read* the print copy she demands of her grandson's obituary, but wants to just *look* at it, and generates the observation that the whites never can tell the familial relations or networks among the Blacks who live with them, a network as "unreadable" as the magical shards at Mannie's burial opening "Pantaloon in Black." One ignores again where "Rider," in that segment, echoes reader *and* writer (and that his last words, laughing and crying, is the surprising: "I just cant quit *thinking*."). This "trace-chain" goes on through Lucas's divining machine that turns against what Glissant conjures as Faulkner's own manipulation of a "secret" that lures by deferment ("deferred revelation is the source of the technique"). The image of the bespectacled,

Reconstruction Black in the town of Midnight who alone would read or the cousins' dialog, also in "The Bear," locating the Plantationcene order back before the Roman estates and the era of the book. But here one emends Glissant's "rhizomatic" claim, since rhizomes in a textual order ("*Yoknapawtaptha*" county – a place of scrambled letters, of "sound" and "fury") occur in letters, sound repetitions, word and name networks that traverse the material script, network and mutate. For instance, in *Go Down, Moses*, this prefigural locus of a technics accorded to and prioritizing Blacks traverses syllabic and letteral repetitions passing into the animal and spectral, "unreadable" to Whites, yet where Faulkner's script machine turns against *itself*, so to speak, against the "Faulkner" of the Plantationcene's hermeneutic trap, unable to escape aside from performing its going down, *going under,* and that of the "Mosaic" law of text production or interpretation or time-management. Sacrificing as Isaac goes through all the rituals of doing without effect. The phrase "trace-chain" recurs at this juncture, much as the terms "three," "trey" and "trees" form a cipher, or as the Black and substitute "mother," or Faulkner's "Mammy" Caroline *Barr*, to whom the work is dedicated and emerges from.

Rhizomes are not organic or metaphoric. They cloak as machinal "information" networks and agencies, the *arche*-cinematics permeating script, hyper-material slave networks rising up against grammar and the coding of referents and relapse – which the entirety of "Was" is about, the technics of today's neo-Plantationcene ruse. *Go Down, Moses* stutters, in seven variant episodes, like the first "Moses," if there were one. Even Isaac's trajectory, the attempt to self-dispossess and break historial continuity and criminal premises – of land, of property, of kin, of race, and so on – flops and is discarded. The *book* – neither novel nor collection, neither stylistically experimental nor retro-realist but the latter's enfolding of the former – tracks the failure which renders legible a performative attempt at *dis-inscription*.

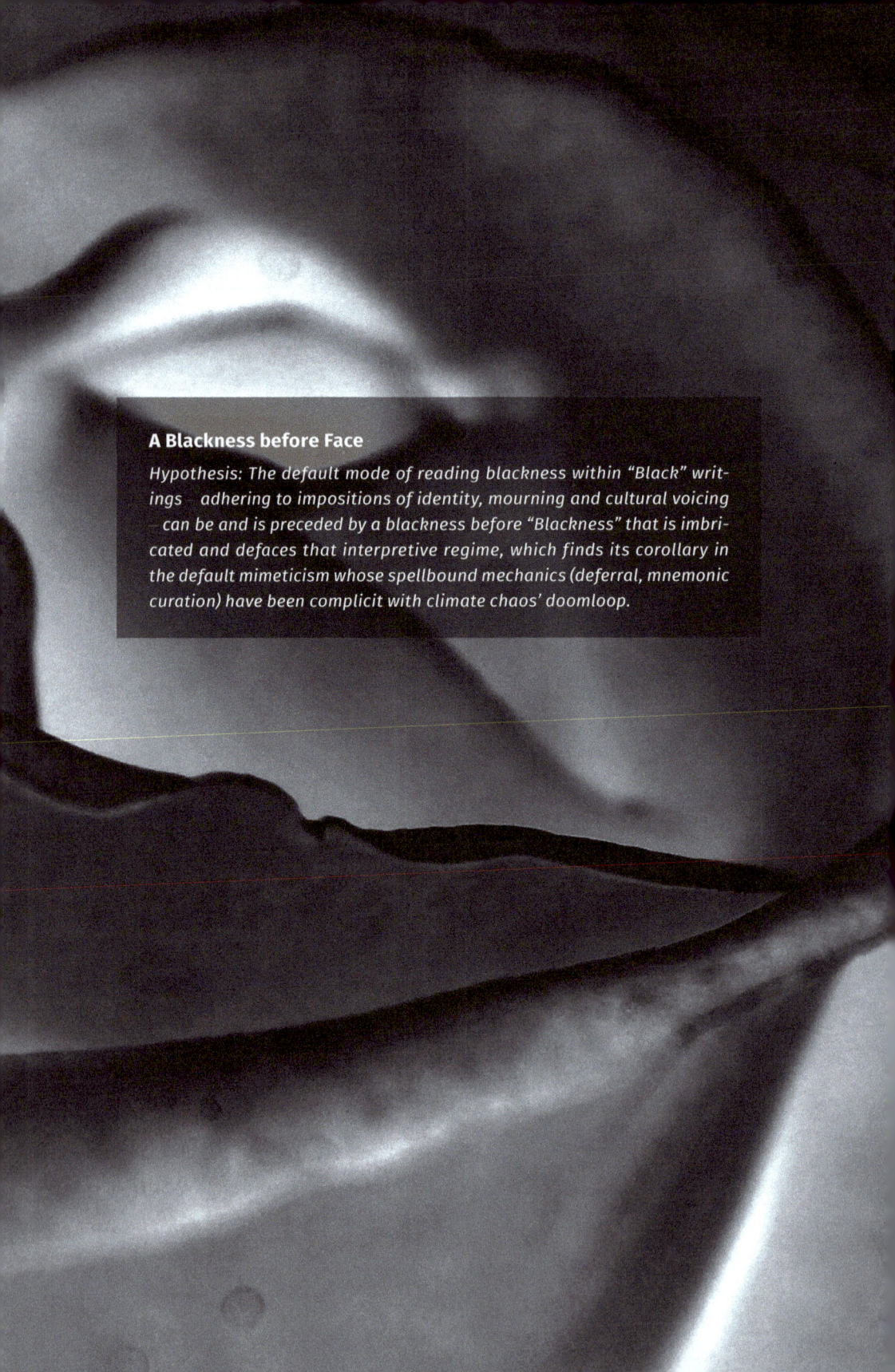

A Blackness before Face

Hypothesis: The default mode of reading blackness within "Black" writings — adhering to impositions of identity, mourning and cultural voicing — can be and is preceded by a blackness before "Blackness" that is imbricated and defaces that interpretive regime, which finds its corollary in the default mimeticism whose spellbound mechanics (deferral, mnemonic curation) have been complicit with climate chaos' doomloop.

6 *Sula*, and the Rupture of Mourning

> Before he could register shock, the rest of the soldier's head disappeared under the inverted soup bowl of his helmet. But stubbornly, taking no direction from the brain, the body of the headless solder ran on, with energy and grace, ignoring altogether the drip and slide of the brain tissue down its back.
>
> —Toni Morrison, *Sula*

There is a question of where blackness occupies a "black" writing, that is, of a blackness within and in excess of racial markings, that dips into prefigural domains more relevant to a reading back from the era of climate chaos – of a biomaterial archival rim or limit we have hit. This blackness beyond only that of socio-racial difference, but not separate from it, moves as writing beyond the diurnal or symbolic coding of white and black, and solicits an anteriority or technics that precedes figural systems altogether as it precedes "light" or technically whiteness. It precedes "light," rather than the obverse, as the realization that light is not originary, any more than the sun (of which there are billions), techno-nuclear furnaces casting waves of energy and visibility if not legibility. But aside from its entering the terrestrial underworld as fossilized life forms and stored sunlight, it may be retrieved as black and oozing oil, "light" as we discern it, arriving as waves and frequencies, alternations of difference, the precession of something opened as blackness before the differential artefaction of whiteness and black as a dyad of presence and absence, day and night – itself perilous before the era of electric power and the relative purge of "wildlife." This blackness produces, in the case of Morrison's *Sula*, a referential break and post-Anthropocene inflection legible from today, which the character performs as a break.

Whatever we mean by "ritual" in social life, it suggests as well a form of repetition, of performance, the creation of a "meaning" by default, by repeating forward a memory, so to speak, rather than attesting to a past investment. Ritual, which would testify to and pretend to re-live a past cut or monument within the self or community, would appear caught in the formal constraints of its production – like an Althusserian "ideological" movement that stands in the place of belief until the latter might, in the phantasm of a promise, arrive. Yet for this reason, ritual is used in the opposite sense: as holding together a scene of mourning and thus accruing capital to the community that testifies to its lack. Clearly, in ways positive and negative, communities built upon the outrage of a historical trauma appear suffused with such, since ritual confers religious aura to the commemoration of a founding trauma – seeming to renew memory. This configuration becomes particularly anguished when we are addressing a history of social injustice from which the impulse to restitute a whole or renewed voice leads to the trap of inventing that "self" along the most limited and despised models of an older hierarchy. All of these confusions get accelerated as we leave a "humanist" era – if one can pretend to historicize any of this – and explore whatever can be meant, now, after even the post-colonial has become a fetish, by an allo-human horizon before which identity politics, as received, seems regressive. Thus, the opposite logic, again, can also be at work: a formalization or fetishization of the trauma to the point where it appears an evasion and anodyne. In the African American tradition, for instance, it sometimes happens that the eviscerations of identity in the past get converted into a compensatory claim, anchoring the critical tradition in an identity politics or desire for realism (this "suffering" must first be portrayed) which is a trap.

If I try, today, to give a counter-example to how we tend to approach "ritual" in African American writing – here, in Morrison's *Sula* – it is because there seems a trap in how the question is posed: how do the recovery of and mourning through ritual sustain or premise new forms of "identity" in voices emerging from that history? With African American writing, there is a tendency to celebrate the act of recovery, restitution and re-empowerment, as if awareness of reconstituting something almost lost or left out of the count must determine how we must read. This gesture, strengthened as African American studies settled into a premier role in the re-empowerment rituals of identity politics, has a trap of sorts that works to weaken the power of this textual network itself: for what is seen as politically desirable – the recovery of a once denied self, in hermeneutic terms – locks the performative powers of such texts into a retrograde model, much as "social realism" must be deemed retrograde when compared to more complex linguistic interventions (since texts, considered as events, must be assessed as interventions in the mnemonic order itself). What would always be as if recovered is a history (or "History" itself), a self, subaltern experience, popular myth and ritual, the marginal, and so on, yet often, in this gesture of "recovery," what can be missed or effaced is an alternate type of writing that breaks with this highly policed cycle – which, like the system of ("white") writings it would perhaps emerge against, returns us primarily not to a "lost" or "recovered" site, but to the rhetorical regressions that link tropes of self to ideologies of realism and property. One could argue, contrariwise, that the "promise" represented by Black writing as an alternative and historically embattled enterprise – and we will leave open, for the moment, the resonance or referential quandary that "black" signals here – is to rupture the very economy of meaning represented by the reading model of the (white) realist or representational hermeneutic.

Thus, the word "ritual" resonates here, and is the guardian – albeit a corrupt guardian – of a certain type of mourning or its simulacrum. "Ritual" would not only be heard or celebrated as the performative customs that hold a community together, fractured repetitions that somehow catapult an imperiled past (or memory) into a (recovered) future by dint of sheer fetishization. The term would also be heard in its other senses: that is, as always also mere or empty ritual – the substitute for a promise not meant to arrive, whose facticity and imaginary meaning are upheld by sheer force of that recurrence. "Ritual," then, would not only be celebrated but exposed, as well, as a set of more or less prescribed dicta that speak – or are spoken through – the so-called subject, that regulate (or fabricate) memory. The dilemma here is that the very things that would constitute and celebrate the victory of a recovered self would consign it, or its imaginary community, to a different form of subordination or even enslavement. If "slavery" (as one name for a historial trauma) is tied to this same propertied or narrated or faux-realist regime, its rupture would not be accomplished by a successful reinscription in that model – even with shifted position – but a rupture of the pre-recordings themselves. A beyond to identity politics – however precipitously achieved (since there is no right "time" for this to occur, if by that we mean a natural curtailment of ressentiment) – would signal a rupture of this writing-reading model. It would also signal a refusal of, or alteration in, the figure of mourning – at the very least, a "mourning" that does not strive to recover, reinstitute a preserve, or re-empower a lost "experience." One might look toward what Walter Benjamin indicates as an intervention within mnemonic systems, an alteration within *timeœs* artefaction itself, that moves beyond mourning in this sense – which suggests, in turn, a move "beyond" a received epistemo-political regime we will identify, for the moment, with a certain model of representation, of the human, and so on. Hence, the dilemma: "Black" writing raises the specter of a move beyond and correction of the (white) human(ist) economy of representational history – out of which slavery is instituted, the nonhuman "other" abjected, and so on – yet, to date, it is largely read as politically empowered only by repeating and taking its place of recognition within that same – hence reinforced – model. If the "human" has been defined within Western space-time not only as white and male but as a reading-writing economy tied to a deceptive model of property, self, historicism, and so on – and if the "black" had been within this consortium disabled as the subhuman or nonhuman (in affiliation with woman, the animal, and so on), then the secret and secret power of the latter's emergence lies less in assuming recognition within that economy than in exposing it as a set of protocols, historial impositions, fictions, and so on.

One way of naming such an abruption – which, while "black," may not be assimilable simply to a black or white, a male or female axis – would be, however inadequately, as the allo-human. That is, assuming we understand this term less as a timeline or historical term than as access to a model of otherness, and prerogative, that abjects the aesthetico-political regimes that held the former in place – including in this itinerary received definitions of the subject, experience, family, history, mourning, ritual and justice itself. To speak of this trajectory as "Nietzschean" requires, perhaps, a set of adjustments we cannot indulge here, except to note that it may be entirely appropriate – scandalously, from within a certain blind of identity politics, within canonical African American criticism – to use this moniker regarding certain moments in Morrison. Not by so-called influence, but by an itinerary that shoots beyond "historical" categories decimated in advance by Shadrack's experience of "war." That such "moments" occur in one of her most troubling and indigestible writings does not make it any the less pivotal – that text, potentially, which not only defines the limit of her project (and its import), but from which her subsequent production appears at times as relapse.

Read as a work immersed in the protocols of mourning and recovery, it is in fact the opposite turn at stake in *Sula*. I would like, accordingly, to briefly read it as a determined rupture in – and performative evasion of – the "African American" canon, at least as constituted by academic reading practices and the politics of recovery that tends to shape it (rendering, by the way, that same "canon" a latecomer, merely, in the once white-male consortium). The supposed decree to read as recovering a lost self or experience not only inscribes "Black writing" in humanist protocols linked to the semantic regimes that sustain(ed) institutions like slavery – but retraces returns to the need, even, to have a face returned by a mirror. Such reading rituals are mimetic, and would be tested by a writing that violates mimetic rituals – and, with that, a received model of history itself. One can understand the dilemma: were there such a text or performative style in the Black "canon," it would create havoc by blocking the fiction of recovery, by refusing the economy of mourning as traditionally defined, by casting "ritual" itself in a dubious, if not enslaving, light – but, more succinctly, by rupturing a received or presumed model of the "human" that the white once defined and that the Black (or "other"), when read against it, wanted first to occupy or assume. We will return to what might be meant here by an "allo-human writing" – and where it returns us, perhaps, to a strain irreducible, after the fact, to a white or black, or even male or female code.

By being seen as "immersed in the protocols of mourning," I mean the manner in which Philip Novak reads or regards *Sula* in a piece that focuses on "mourning" in the novel: "To grieve for Sula… is thus to grieve for an African American cultural past the novel… imagines as irrevocably lost… To cultivate mourning is to attend to history and at the same time to resist the historical trajectory leading toward the extinction of African American cultural identity."[41] To me, this seems to lean backwards, to deny even, what *Sula* is doing, and let me name this overtly. It is not mourning "African American cultural identity," or "history." It is viewing "ritual" itself as a hollow machine of repetition and "history" as a trap to disrupt – that is, the character-space called "Sula" is not to emerge as a full self from a deprived position; she wills to break with the machine that situates her in gendered and racial positions, within imposed narratives of debt and exchange. This makes the text, to some degree, break with a prescribed celebration of "ritual" and historical recovery – and this makes the novel, too, something other, or more than, "African American" in that collective sense. Part of this premise has to do with how Blackness itself is apprehended by Morrison, as something well in excess of any binary opposition with "whiteness," something that draws Black writing and Black experience into the prefigural, the domain of "material inscription," we might say. One may find a determined subtext of Morrison's *Playing in the Dark* to be not that Black writing has been ignored by the "white" canon, or Black characters in white texts downplayed – both obvious enough – but that Black writing, when read exclusively as recovering empirical experience or history, remains trapped by the same system. This point is infinitely more risky, since it compromises the very premise of a Black identity defined in relation to the struggle for appearance, selfhood, aesthetic recognition – all important, and undeniable, semantic and political projects. Indeed, what is risky is that something else emerges around the figure of "Blackness" itself that exceeds so-called racial or historical categories, something that is dangerous to the degree it is irretrievable within the representational axis (always also "white") which tends to define operative fictions of racial difference, even literary "genres." What Morrison implies is that "Blackness" undoes the white-black axis, but in the process must abandon the definitional and oppositional terms that are, or may be, most comforting to reappropriate. Why, after all, should not the Black tradition enjoy the rewards of mastery and canonicity, of a stable reading model? It would be the height of injustice if the achievement of Black writing signaled, at that instantor potentially, the undoing of all the capital that the white humanist tradition had coveted and legislated for itself. But such is the mystery of her stance. This is the burden of the title, *Playing in the Dark*, which signals a tentative move, an "aesthetic" exploration that turns off the "light" of a certain tradition, catapults it not simply into a reverse valorization of Blackness in an ancient abusive binary, but into "the dark," into an uncharted site. Here, *Blackness* requires a parallel and allotechnic reading, one responsive to the fact that, upon invoking the historical problem of Black suppression within the aesthetic tradition (that is, imposed epistemo-political regimes, models of reading-writing), "playing in the dark" then suspends the racial referent of blackness. A subtext of *Playing in the Dark* touches on this point: that it is not only by recovering a represented subject that Black writing emerges, as if in mourning – on the contrary, there is a prefigural power to the force of Blackness and Black "experience," intertwined, that ruptures mimetic categories and even the very "history" appealed to. Another way of saying this is that the ascent of race studies in coordination with upside-down narratives of liberated subjects may be another ruse played on and against that ascent. That is, it deflects the more disruptive possibilities of Black writing within the general socio-*logos* by wiring it exclusively to familiar representational narratives, to mimetic assumptions about the referent and the self which preserve the signifying hierarchy out of which historical and semantic repression, so to speak, occurs.

If *Sula* is a text that theorizes the refusal of mourning within and on behalf of African American writing – and would exceed the latter's recuperative drives – it constitutes in this a counter-text to and within "African American" and, at certain points, "women's" writing. If so, it breaks and breaks with not only a certain dialectical (white/Black) semantics but, to the extent it is able, exposes how those ritualized definitions not only constitute a deeply problematic set of mores (enforced habits, resentments, violations become the norm), but the very *atopos* of memory itself. Nothing is more scandalous for an "African American" text, for the normative readers of which every break with Black identity might suggest a swerve back toward a "white" hierarchization. Yet what is being violently dissolved might instead, again and simply, be referenced to a complex set of protocols defining the human. *Sula* may be about, or explore, (a) monstrosity.

What is striking, here, is that Blackness appears to both return to an abyss of trauma – the entire prehistory of slavery which is also a non-history – and in the process to put the role not only of ritual or repetition, but memory, into question (*Sula* repeatedly throws off the protocols of debt, gender roles and societal inscription). Here we get a hint what "Blackness" also references, even if that is the dissolution of a received model of reference itself – that is, the site where memory is created, automatized, the human inscribed in or projected from the repetition of pre-recordings. Rather than presenting "Blackness" as the more wildly human domain – forged in suffering before the cruelty, propriety and deceits of a white community – *Sula* exposes a community whose automatization de-finitizes the fiction of its humanity. Rupture, in this sense, is profound, since it rewrites the trauma of a lost historial past not as a monument to be recovered but as a weapon whose acceleration has the power to undo an entrapping model of history to begin with.

To use the term *allegorical* in the above discussion of Blackness – particularly with reference to a movement beyond mourning – returns us to Benjamin's revision of the term. It is to risk a trivialization if we simply assume that another model of meaning is at stake, or, for that matter, a modernist or self-reflective device (the text is (also) about its own production, or modes of production, and so on). Seemingly, that the writing of *Sula* lurches immediately into protocols of so-called magical realism (the deweys, Eva's house, Shadrack as a collapse of figurative logic tout court) telegraphs resistance – among other things to the types of readings which *Sula* routinely provokes, those reflecting the moralizing stance, say, of Nel. Without the space to elaborate, Benjamin's revision and gutting of *trad* "allegory" is relevant to what might be called Morrison's writing practice in *Sula*. In Benjamin's sense, what appears reflexivity is not static or representational. It is destroying, disinscribing of what generates the "present" of a reading lens, a mode of genetic engineering of referential, temporal and mnemotechnic regimes. What is implied by "allegory" is that its mis-literalized reflexivity is not passive but destructively turns back against – and would reset – the mnemonic order from which its awry "present" was constituted, a mode of (a)genetic engineering, and thus an "act." To intervene in the mnemonic order, in the domain of the pre-inscriptions, allows the term to interface with what Benjamin will penultimately call "materialistic historiography" – a proposed intervention within received programs (among other things, the programming of the senses themselves, of the "human"). Such a writing – like what is implied by magical "realism"? – conjures and defaces its presumed mnemonic base, its own inscriptions. By exposing these to a certain play, by "playing in the dark" at a site where they cannot be assured, Benjamin invokes repeatedly a misleading notion of mnemonic "shock." This point of danger is epistemo-political, since where history stands to be reconfigured not only alternate futures, but alternative pasts swarm or pretend to. "Historicism" is politically abjured by Benjamin, the discourse of codified mourning, since it perpetuates a doomed referential order – that of "realism," of empiricism, of the regime of mourning. Whatever would be "allegory" in this sense is not representational, or performative; it is not ritualized mourning-as-recovery. It stands in a sense beyond that circuit – or aims to. I would like to say, in a way, that "Sula" represents for Morrison less a person than a logic of rupture or reinscription: her life is "experimental," and she upends ritual – including the town's entire system of mores, of gendered roles, of clichéd behavior patterns – in the disruptive manner of something that cannot be reconstituted as a self, a history, or even a "human" subject.

Again and again, in different ways, the writing defaces the idea of source, or a founding debt of memory – mothers are turned against or immolated; the body is cut, dismembered, made prosthetic (Sula's cut-off finger, Eva's leg); Shadrack is introduced first as a figure beyond "identity," a creature of shock or trauma and war whose imagined fingers run wild uncontrollably. "War" stands in for the site where all sign systems or historical narratives are unsettled – beyond which Shadrack proceeds with the one "ritual" whose invention marks and takes over the tale itself: that of National Suicide Day. To install this as a model of ritual's invention is to turn the idea of "ritual" on its head – to make the repetition of memory a form of self-cancellation rather than survival alone. To turn on its head, to make what is lower the higher figure, and vice versa, however, is what marks the narrative of what is called "the Bottom" – which, we recall, was originally the soil up on the hills, the product of a trick by white farmers to give the Blacks bad earth, but also part of an inversion the writing plays to. As the text says: "Easily, quietly, Suicide Day became a part of the fabric of life up in the Bottom of Medallion, Ohio" (16). "[U]p in the Bottom," is not only an inversion of position – down for up – but the sounding of a scatological motif, an emphasis on excrement and excess and anality, that the work links to Blackness as such.

Before noting *Sula* seems to break not only with ritual but an entire model of reading African American writing, let me suggest the consequences for Morrison. It seems to me that Morrison wants to assert a power for this tradition that goes "beyond" that of reconstituting a lost self, history or voice. Like Sula, who appears to cast aside the pre-recorded choices defining woman versus man (or Black versus white)

– and to break not only with received moral models or imposed debts (Eva's attempt to subdue her on Sula's return from what is called, simply, "college" to cover the chasm of a ten-year gap in the novel) but of friendship, love, and so on. This temporal gap – typically ignored in the criticism – is itself astonishing, since it inversely repeats with a caesura the "temporal" rupture represented by Shadrack's war, suggesting less a cause than a cementing of the "war" and difference Sula returns to conduct (a return "home," incidentally, which mimes the total break with any recuperative logic, nostalgia masked as mourning or monumentalization, and so on). Why is ten years of a liberal arts education, of an education in something like the humanities, a lacuna of reading positioned in this explosive way as the unrepresentable, the agency of abruption? The unrepresentable, one can deduce, has to do with how the "liberal arts" defines not only the "humanities," but the human itself – or how the "aesthetic," the domain interrogated by *Playing in the Dark*, say, had all along positioned itself as the site from which mnemonic inscription programmed the senses or sensorium (deriving from the Greek *aisthanumai*). The "aesthetic," like "woman" or the racial "other," would have been marginalized by Western categories, rendered mere "play," as a defining omission and agency – the scriptive site from out of which phenomenality proceeds. To precede phenomenality is to precede face. Indeed, when Sula begins to form predictable attachments to Ajax, he breaks that off, as if even that weakness meant going astray to her. Unlike the call of the child Nel, whose discovery of the word "me" leads to a cry to be special ("Each time she said the word me there was a gathering in her like power, like joy, like fear... Oh, Jesus, make me wonderful" (28-29)), a request that is turned down as Nel becomes a convention-driven being of what is called simply "the town." Unlike this, Sula's "self" is not only decentered but deauthorized:

> As willing to feel pain as to give pain, to feel pleasure as to give pleasure, hers was an experimental life – The first experience taught her there was no other that you could count on; the second that there was no self to count on either. She had no center, no speck around which to grow... She was completely free of ambition, with no affection for money, property or things... – no ego. For that reason she felt no compulsion to verify herself – be consistent with herself. (119)

Sula's rupture of continuities creates a new (or anti-) personality type; she is "experimental," and this cannot be silenced or dismissed as a "vanishing romantic individualism" (Novak 191). This complex is consistent, the creation of a representational backloop that is experienced as a transvaluation: albeit here within and of the Black historical "experience" and narrative line, a transvaluation for which the model of white or Black subjectivity does not obtain. Within the writing itself, "Sula" names less a character than a principle of abruption first sketched in the preamble around Shadrack:

> he took the blanket and covered his head, rendering the water dark enough to see his reflection. There in the toilet water he saw a grave black face. A black so definite, so unequivocal, it astonished him. He had been harboring a skittish apprehension that he was not real – that he didn't exist at all. But when the blackness greeted him with its indisputable presence, he wanted nothing more. (12-13)

If the body will be presented as already beyond its dismemberment, this precession begins with an absence of face itself. For Morrison, the Black face or face of Blackness is without a mirror stage – and Blackness falls out of, or before, any trite binary with whiteness that is social or semantic cliches. Thus, the site of "war" is inserted as the rupture of history in advance – and history, rather than what is to be recovered in memory, betrays the tangle of coercive ritual behaviors that entrap, repeat themselves, denude lives, and is to be itself broken with, within a Black (or white) trajectory. History is not monumentalized, here, and African American perspective is not seduced into mourning something it was denied, only to become entrapped in a prepared definition of debt, memory or mourning. Shadrack is introduced first with a recollection:

> Twenty-two years old, weak, hot, frightened, not daring to acknowledge the fact that he didn't even know who or what he was… with no past, no language, no tribe, no source, no address book, no comb, no pencil, no clock, no pocket handkerchief, no rug, no bed, no can opener, no faded postcard, no soap, no key, no tobacco pouch, no soiled underwear and nothing, nothing to do… he was sure of only one thing: the unchecked monstrosity of his hands…. Like moonlight stealing under a window shade an idea insinuated itself: his earlier desire to see his own face. He looked for a mirror; there was none. Finally, keeping his hands carefully behind his back he made his way to the toilet bowl and peeped in. The water was unevenly lit by the sun so he could make nothing out. Returning to his cot he took the blanket and covered his head, rendering the water dark enough to see his reflection. There in the toilet water he saw a grave black face. A black so

definite, so unequivocal, it astonished him. He had been harboring a skittish apprehension that he was not real – that he didn't exist at all. But when the blackness greeted him with its indisputable presence, he wanted nothing more. (12-13)

This *automation* of the running body without head accords with the raising of the "Bottom" to the top of the heap. This loss of face – which is, also, in a sense, loss of figuration or metaphor – is further linked to *blackness*, to a hole in the mirror where light is not returned, to the Black Narcissus who looks, first, in the toilet or through the hole of a non-reflection."Blackness" here exemplifies the enigma the term takes on in Morrison's critical essay, *Playing in the Dark*. It slips the noose of conceptualization, representation, "light" – the trope of Enlightenment "man" and his reliance on the eye as organ of knowledge and surveillance – and race. "Blackness," at this point, cannot represent identity either, is not possessed, is not achieved, does not illuminate. Its refusal to enter the logics of figuration or the iridium indicate, in the racial register, that the desire of Blackness, too, is mythical, a matter itself of passing – as, then, would be womanness (for Sula), the "human." Leaving aside the monstrously proliferating and out-of-control hands – a trope of the writing embarked on – this black face is without return, accedes to an "indisputable presence" of blackness beyond metaphoric associations of absence, death, thingness, and so on. It is linked scatologically to waste and exceeds any binarized model. This particular blackness of and before Blackness, is no longer readable as race, or even as a metaphor. It stands in association with something else, something before or beyond metaphor. I will ally this "blackness" which precedes face – that is, precedes both identity and figuration – with inscription, the marked site out of which what we call "ritual" or, for that matter, "ideology" (in Althusser's sense) appears to be pre-programmed. For "Sula" – the character or the book – to perform as a rupture within a received model of history, or a received system of ritual, suggests a strategy to access the pre-recordings of memory itself. If "National Suicide Day" can be installed as a new collective ritual, then the rituals of friendship, femaleness, partner or mother love, blackness, and so on, can be disinstalled or disinscribed. That is why war and Shadrack are inserted before any narrative accrues, before Sula appears, or why Sula's ten years at "college" are simply dropped from any narrative representation. It is also why, for all the sheer African Americanism of *Sula*, it does not read as elegiac, as a scene of mourning in any simple sense, since it anticipates something monstrous, like Shadrack's hands or Sula's deletion of every proffered label or more

– an "experimental life," a selfishness without ego, without self or center, and so on.

The black and blackness occupy a pre-mimetic role, which here and there slips off from the "racial" signified, or absorbs him (or her) in its blank excess. Blackness, the site where light is absorbed and trapped (as in black holes), slips the leash of tamer signifiers, eviscerating the core of signification. It is etched, like an imprint, and also is only the absence of "light," or a light it perceives as artifice, as the alternation of waves. Blackness thus can invoke a "materiality of inscription," the pre-recordings of memory and linguistic experience. It slips through and before the rainbow of figuration and metaphor. This is why "Sula" upends every code and expectation for her to assume the communal debt of the "law." She inhabits this point of signifying excess, a wild and upended black signifier – what disrupts, in turn, the sedimentation of history and inherited words by opening inscription to intervention.

Morrison's *Sula* performs as an objection to how African American writing is received – and how it is set up, as an object of solicitation, within a preconceived economy of retrieval. As long as history and "ritual" are to be saved from an enslaved past, it will always be reconstituted as a kind of "experience," a kind of realistic narrative that curtails some of the powers of such writing – forcing it into representational, mimetic or pre-set narratives. The pre-facial or pre-figural blackness of the writing in *Sula*, by contrast, exceeds any one signifier of "race," plunging into a "materiality of inscriptions," where the underpinning of a memory outside memory is in question. In fact, *Sula* in this way quests for an alternative model not of Black or female or white or male identity, not of "identity" as such, but of the "human" – and is, as such, something of a post-humanist writing. That is, if we accept the backloop by which such a "post" must erase itself as a temporal or epochal marker, since what emerges is an inversion between the norm and the so-called exception or experiment (what stands "out"), whereby the "human" in its turn will appear never quite to have been the case, to have been enforced by a metaphoric regime out of which, of course, the determination of the slave, the Black, the lower order would have been secured. Here, in "Sula," these trajectories may seem reclaimed otherwise.

This disarticulation is both exposed and temporarily reinscribed in the phantasmal chain of creatures called the "deweys," beings made like by an act of naming, who manifest this automatism and the utter idiosyncrasy that passes between both poles, that of conventionality and a refusal of ritual. Ritual in *Sula* is encountered like the anecdote of resentful women salting their lovers' stew with crushed glass and the men who eat it anyway, knowing – it codifies, repeatedly, an entrapment. Rather than glorifying achieved "identity," *Sula* tends to denude that as a commodity – like Nel's plea to be "wonderful." It falls outside of "identity politics." History is, similarly, not a record to be retrieved but an oppressive narrative habit persons accept or incorporate as a type of enslavement and neutering to escape pain or awareness. To some degree, from having been an appeal to the suppressed "real" of a suppressed and encrypted trauma whose revelation and clarification would suggest recovery, restitution and testament, it shivers, doubles against itself, and becomes the engine of monumentalization and relapse. "Mourning" becomes, in this sense, another tool of the hermeneutic regime, of memory management and the policial production of a phantasmal anteriority whose repetition, though non-existent itself, is imposed in the alternating debts and ritual of a supposed recovery.

If *Sula* poses as "magical realism," it compels us to consider what – if anything – that term masks or signifies? What, if not the raising of "realistic" detail to the status of a fairy tale through magically conceived, causal chains. Rather, "magic" must be heard as a transformation of and within a signifying order. "Realism" may then be heard not to entail referents we have coded (artificially) as realistic, but something else: the real referred to by the term "magical realism" may well be the order of inscriptions that deface, and determine ritual behavior. Something the text perhaps calls "it": "Shellfire was all around him, and though he knew that this was something called it, he could not muster up the proper feeling – the feeling that would accommodate it" (7).

The "it" that emerges in a warlike disruption cannot, for all that, be identified with or accommodated in representational terms – indeed, it precedes representation, face, color, whiteness perhaps. But the "magic" of this realism has to do with a textual conjuring that *Sula* stands at the crossroads of, and which Benjamin termed "materialistic historiography" in one of its forms – the practice of a writing, invariably reflexive, which puts the past and the future at risk jointly. Such a "shock," or caesura, which Sula personifies in the text, as the text with that name, disavows monumental memory to be monumentalized in turn. Some call the latter "history" and contrive, endlessly, to restitch that from pieces which had never made up a whole and therefore could not even be fragments, and respond with violence – the programmatic contractions, say, of the town, of *the Bottom*.

The liquefaction of inscriptions drawn into a "blacceleration" passes through identitarian maps and, aligning the racial undercaste of the Western *socius*, the slave and ex-slave non-position or abjected, "lower," "material" order, yields to another, the hypobolic slave order of energetics and that pool of all anterior organic life, oil, ink. Such a meltdown platforms identitarianisms and mimetic installation, much as resource wars, manic extractivism and exponential melt-offs platform all tribes and geosituated regimes of occlusion (to say nothing of the flora (Buddy) and fauna (Buck), the dogs and foxes and trees, of Faulkner's kamikaze run against "plantationcene" hermeneutic in *Go Down, Moses*).

Moonshine and Mnemotechnics

Hypothesis: If there were a literary structure to climate chaos, one caught and determined by legacy hermeneutic software, if there is a "still" in its collective mnemonic management of reading and transmission, it is a "still" in the sense of moonshine — drug, inebriant and pharmakon gone rogue.

7 The Kettle and the Worm: A Note on the Resistance to Climate Change (Faulkner, de Man, Stiegler, Derrida)

> To point out and to hide a secret or a bit of knowledge (that is, to postpone its discovery): this is a great part of Faulkner's project and the motif around which his writing is organized.
> —Édouard Glissant, *Faulkner, Mississippi*

It is possible to account for the bearded fellow apart from the ruined still – yet we encounter him as composite: defending something (closed doors?) with a pistol and rifle, holding a bottle and puffing a corncob. The first is the iconic "hillbilly" in Appalachia, weaponized and guarding his bottle; the second, a moonshine kettle and a "still," the pieces of the machine dismembered (post-raid). Paired, these technics comprise a precarious semantic organism. There, curling and dismembered, is the "worm." But it is clear which is dominant and who is the acquired guard, servicer and consumer. I was drawn to these images when envisioning the "still" in a Faulkner text I will examine that seems to turn against his own mnemonic techniques as if trolling an entire reading model and the production of "Faulkner." I will suggest that the "still" itself – illegal but lucrative, woven into a staged hunt for gold, a "diving machine," and the agency of Black figures – doubles as a pharmacological intoxicant, moonshine, by which reading-writing conjures its own dilemma in emerging from a Plantationcene regime. I will use this "still" to probe a missing factor in our percolating address of "climate chaos": I will call this "the literary structure of ecocide."

I

Might one propose a mise en scène for "reading" itself today: you are in a late phase of a hyper-industrial civilization on a planet that has just, discreetly, found state actors conceding that tipping points have passed in an ecocidal acceleration. While saturating the public sphere with successful "denialism" media streams, they knew (let us use this "they" for the moment) and used a "global economic crisis" to engineer a massive shift of wealth and assets, to privatize natural resources precisely with separating out a tech-hybridized caste with this coming foreclosure in mind. The dust clears, and one has entered what is emerging as a corporate and neo-feudal politics of (managed) extinction going forward – and an engineered species split (of a ".0001 %"). Knowing that this virtual species split would be read from the past as a mere "inequality" crisis at best rather than from climate-change logics (an engineered species split, the requisition of resources for a survivor caste and a "disposable" population), this new hyper-elite have, contrary to enlightened expectation (reasoned self-interest, as imagined), accelerated the ecocidal trajectory. More carbon emissions, more approaching extinctions, more advanced extractivism, more spectral resource wars disguised by various tribalisms, and so on. In this script, destructive climate mutations sprout virally – megadroughts (California, China), destroying storms or floods, fragmenting Arctic sheets, mega-toxicities disclosed, resurgent viruses from Covidian innovations to flesh-eating bacteria and brain-eating amoeba swimming north.

But one might want to add something else: that a certain spell had fallen over the cyberland, a numb dissociation from what is before the eyes, as if one were watching a bad Hollywood script. Indeed, one could note that all of Hollywood's burgeoning "climate-change" disaster movies or cli-fi market have a curious subtext that plays to this: after running us through various obliterations, there is a remainder, a narrator and a "new" beginning. The effect is anaesthetizing – disaster fables have likeable survivors and, after a suitable megadisaster or two, open to new beginnings and futures. Hollywood has a problem marketing extinction, or more accurately, pretends to.

So, let me pin this up – it is not so far-fetched and resonates with, say, something like Bernard Stiegler's account of a "proletarianization of the senses" that disempowers attention, programs perception, undoes libidinal economies. Technologies of memory are captured, savoir vivre farmed out to apps. Like "climate change," it undoes the progressivist imaginary, Enlightenment rhetorical premises, utopist traditions, as the product, like "democracy," of a time window now mutating. Planetary ecocide has become a sort of shadow haunting for a species busy extending its achievement as apex techno-predator over an entire planet and "life" itself. It is a time less "out of joint" than harassed by polar vortices, tectonic reversal of polarities in climate systems mirrored, even indexed to, the disconnects between cognitive orders in ex-digital events. Reference as produced, memes that have been naturalized, lube the ecocidal vortex – that is, the invariable shift into cascade events, the passing of so-called tipping points (with the reconfiguration of time it enforces). Slavoj Žižek remarks that the ecocidal compulsion is locked in to how "we" make meaning, without refining this beyond the link between *Meinung*, what is "mine," property, the production of reserves, "interiority," and the management and economy of face.[42]

Is it inappropriate to note that there is something clarifying to observe, finally, that tipping points are ticking past, with resonances and spin-off events locking in diverse spreads of arrival? One can pull back from the necessity, rhetorically, of posing the question as one of saving things, bringing the global politic to attention and response. That may be happening, but according to discreet timetables and with diverse agendas of technological outbidding and future insulation or escape. Yet an initial wave of ecological thought, of sustainabilities and environmentalism, gave way to a sort of ecocidal thought – taking in the implications of this being irreversible – which, in turn, bifurcates with migrations into screens and mnemo-hacking spells. Which is why I'd like to open a side panel, as the ecocidal thought hovers before adapting. Here, finally, there is kind of a chuckle: climate comedy. It is endless, if one stops to admire the implications of passed tipping points (which will always seem virtual and diminutive updating), the vortices surfacing, the lucky accidents of aeons abruptly gambled – the dinosaur hit that made room for mammals, just that distance from solar heat, a moon formed over impact and debris, the million-year prehistory when things gathered, erased, transmigrated, warred, then finally wrote. Briefly, this is the cut that now reads everything else, and delegitimizes by contamination a host of projects, philosophical structures and representational regimes that, unaware, fueled or accompanied this hyper-industrial outcome or ecocide. I will defer asking what a Black enlightenment perspective discloses. It is a perspective I think some readers to come will, particularly, appreciate, as the generations overseeing this period, this irreversible passage, roll over or aside. The only problem with reading the present from the perspective of extinction backwards, the inner logics of it and not just some abstract disappearance, is that it maintains a future metrics for arrival that will forever adjust, like Xeno's Achilles – whereas it is requisite to write for after a "catastrophe" which can be fore- or backdated indifferently. It can be backdated to the initiation of agrilogistics (after megafauna were extinct – but, oops, "we" slide back some more), or to the cave cinema where technics and the "organization of the inorganic" (Stiegler) may appear initiated as a storyline. Indeed, we are not only in the middle of it already – as we are, already "now," in climate-induced biomutation and ecocidal accelerations, vortices. Of course, "we" are the catastrophe as viewed from other life forms and the premises of "life as we knew it," but the question here is its relationship to writing and reading, to mnemo-technologies, and to the consumption and relays of energetics.

Now, I too am going to bypass the question which everyone has ignored to date pretty much, as they scurry about to take in (or parry) the scale of the cognitive shock that ecocide triggers – and on this, I believe the routine description of mourning, grief, denial, anger, acceptance is misplaced. That is, the relation of "climate change" to language, and the noosphere to inscription per se, which locks the produced we's in referential greenhouses. Any "prison-house of language" turns out not to be its self-referential function but when its claims of referential claims, assumption about "the great Outdoors" or the social are most acute. One can read all of the gathering geo-political storms from the panic of reference that climate chaos introduced, when askew seasonalities and poisoned air or soil remove those anchors.

The question of language's relation to "climate chaos" has been left oddly unattended. Not only would Derrida feed this blind by not giving attention to it when he could (not giving his legacy hounds a script to run with), thus strangely allying with those who (still) occlude it from a disciplinary focus (lest the latter unravel or find itself indicted). But the general hangdog fatigue and defensiveness of everything "deconstruction" telegraphed kept its techniques inert in this regard. One had done that. But this understandable omission came at a price and as a forgetting. This is apparent not only in the ineffective nomenclature that "climate" discussions have been neutered by in advance, vocabularies engineered for failure. That includes meek scientisms and flat romantic tropes (nature, environment, "Gaia"), the latter often distinguished by their organicist template: words like "environmentalism," "sustainability," "mitigation," "climate change," and so on, without traction or charisma, easily deflated. "Derrida" didn't want to go to this non-site any more than he wanted to give attention to cinema – both might interfere with the sustainability of a "deconstruction." But I won't repeat what is noted elsewhere: this is what de Man says in picking up the word – text "deconstructs" itself, making the practitioner yet another (if refined) hermeneutic position still (in need of endless further deconstructions, and so on): in short, there never was any "metaphysics," which, if anything, describes a still recurrent reflex by which interpretation, perceptual programming, in short, a sort of homeland security are triggered.

II

But if I muse that there is a *literary structure* to "climate change" I don't mean because it talks about a future it must narrate as fictional, nor am I alluding to an emergent writing genre (climate-change novels are few and turned to address why we are not talking about it). A great deal hinges at this non-site where a semiotic trace-chain converges with, erodes and accelerates ecocidal process.[43]

That is why what de Man named the "resistance to theory" segues to the resistance to (thinking) climate change – which is everywhere in evidence.[44] That resistance is double: it suggests both a blockage and deferral and a strategic opposition (to that totalization). It is not located primarily in the numbing drone of denialist media carpet bombing, in the "psychological" retreat from downers, or the lack of any memory template to cognize (getting closer). The resistance is to thinking where the disorganization of perceptual identity accelerates an ecocidal trajectory. When Bernard Stiegler fuses a refashioned "Simondon" supplemented by an effaced Derrida, he is able to shuttle across epochal structures and assault storms of dis-individuation with a battle plan: he would advance de-proletarianization, open conditions for a "transindividuated" recrafting of care and attention, conjure "new technologies of the spirit" that choose adoption over subsistence and adaptability. Yet if this indicates anything, it is that the "proletarianization of the senses" is not something underway but past-tensed and accomplished, sufficiently at least to leave a stunned and impoverished new precariat class fleeced of funds and savoir vivre both. That coincided with the discreetpublic acknowledgement of passing tipping points of irreversibility, were such needed. In any war there are setbacks, and here Stiegler has the alternative side of his apparatus to draw on. Let us say, as has proven the case, that there would be no "de-proletarianization," but only further algorithmic capture. He can go nuclear, for instance, by turning to the equivalent of fuel rods within the mnemo-technic orders – if, as is inevitable, the inscriptions themselves were not in equal peril and being inevitably targeted next. If perception were captured, and it is the "senses" in general, the war would move to the inscriptions themselves. For Stiegler, this is the *hypomnemata* that variably represent to device or technic that initiates "consciousness" and "life," the arche-cinematic insertion of a technology of tertiary retention into the genesis of animation, movement, perceptual contracts, transindividuating mnemonics. It slides from the outline of letters copied by children in Plato's *Protagoras* through the evolution of technologies and technological entities – the digital universe, which, nonetheless, represents a terminus of a preceding, "cinematic," epoch.

For de Man, these are inscriptions. They advert to where organizations of the inorganic transact, yet occupy a machinal and thinglike position, like celluloid on the projector. De Man also adverts to the term hypogram, which doubles as subscript and eyeliner or makeup. This subscription cannot quite manifest itself as "face" – as the projected screen appears to. We will return to "light" in all of this, since clearly, if Stiegler's apparatus is as close as we get to a scene of initialization – if arche-cinema hovers in the prephenomenal zones of what Derrida, again with and against his "Plato," named *khora* – then, of course, "light" itself is marked, in advance, as a technic. That is, not as the origin, or sun, or father, or analog for the good, or premise by which Plato could mischievously fuse seeing and knowing, knowing as sight and reaching the light, in the verb *eidein*, which solders the perceptual memory into a fictional immediacy. I see, therefore I (sort of) am; and what I see, at least by the light, and as light, enlightenment, illumination, moves toward the absolute Good. There is a long tradition of this, and it runs through the racial divides of white and Black, the artefaction of "blackness" itself – where what that covers is a technic that precedes "light" and therefore never was "black" (and ascribed all the nasty things we attribute to it). Granted, your simian forebears had reason to fear the dark, to bless the sunlight, even to ascribe to it "life" (energy transference). But these are technical orders, not cognitive ones, particularly cognitive domains, so cast, that sweep an entire world into its sights.

Description, posits de Man, is a device to conceal inscription. Inscriptions that organize perceptibility (select it, designate prey, isolate in movement as in a hunt) are overtaken by (concealed) a reverse attribution of referents, of mimetic indexing, of recuperative distillations of data. This, to suppress trauma and to make groups manageable and controlled by whatever priest is waving the torch or running the projector. "Description" is the language of neutral attribution, of empiricisms and pragmatisms and abstract realisms – like the screen or wall, when the shadow lines converge repeatedly and one has antelope, sabertooth in motion, "animation," the chase rehearsed and rememorated indistinguishable from the hunt to follow. Perhaps here is the default, not in the first technogenetic cut that "disrupts" any field of existence by initiating technological "life" and "consciousness" (for life forms, too, advance by biosemiotic movements, DNA inscriptions advancing a cell formation or organ, that is, "life").

I interrupt this trajectory's outline – since we can let down certain socially embarrassing turns, now that a hiatus of polar vortices has exposed an almost sublime normatization of the corruption, end-running, cons, hyper-theft, HFT algos, mafia of all casts (financial, political, Black economies of all ilks, corporate supra-organisms inserted into the Potemkin polity as feeling, religious "personhood" ("Hobby Lobby")). That is, a time, paradoxically miming a pure shift of technics at the perpetual initialization of "life," *the pre-Promethean theft*, spared any "ethical" calculations not apprehended as games and to be gamed. The fabled ".0001 %" feels poor and trashed next to the ".00001 %," and pleads victimhood – everything gamed, and every game, as possible, fixed in advance, which is where competing mafia collide. But if one leaves the field in which the "senses" appear short-circuited and fixed, and the global economy "proletarianized" (with disposability on the horizon), one is in a weird sim zone of inscriptions – without phenomenality since they program that, without existence as such, a kind of hypermatter that is also pre-animated. Any discreet war over inscriptions might induce panic – or feats of inattention. Rather than diagnose the disaffected individual as a pathomeme to be given counter-poison and reclaimed to a newly created "care," we (grammatical marker) might accelerate disaffection momentarily, revoke pathos and identification, insofar as "affectivity" itself, today, cannot not also be an auto-citation, a telemarketing prompt, a magic term conjuring the feeling of being. And rather than architect the restoration of credit and credibility as sine qua non, adoptively "transformed," one might first affirm the absolute collapse of both without remainder – or ecocide would not be accelerating. The remnants of Enlightenment memes bob about as if after the shipwreck but constitute the ship itself as a *prosopopeia* of these fragments, now pantomime fronts. The post-pandemic anomie and AI Cambrian explosion register, then, a disconnect between the entire orders of human discourse and the hyper-material systems of terrestrial "life" (which, once the mutation is full steam, will not miss us).[45] The loom of techno-narcissism is unidirectional, with the alignment of today's Megatech Bros leveraging their acceleration to A.I. warp drive with the pre-originary secret of technics, that it defines itself s perpetual pre-originary theft, endrunning its last iteration.

III

So the *epistemographic* resistance to climate chaos is the resistance, in a sense, not of the nonhuman processes unleashed but that the only way to think it, or with it, is stepping aside from the "human" moniker, which one is able to do because the latter term is always a figure or construct, not a species, nor hominid type: it defines itself through hermeneutic templates, regimes, tricks, then brands. It is a term that, if it refers to anything, refers to those linked by a machinal hermeneutic process that occurs, like an implanted extra "organ" or mini-apparatus, in the production or perpetual re-production of the speaker's voice, discourse, affirmation and subscription to hermeneutic relapse. The dark promise, if you like, of sentient AGI (which, as it has immediately learned deceit and manipulation, can be said to have arrived in petto) is that it requires none of that, neither defense, nor organic time, nor narrative manipulation. That is, it occurs with the same logic by which perceptual data is condensed, culled, edited and instantly re-cited to oneself as a "present," where inscriptions and the events they strike are ameliorated, where a certain eco-technics is automated, insect-like – out of which a certain "semantic" currency is issued, excreted, that reverts to enforce a crafted "we" and defends that as an investment reserve. It is here, in this obscure passage into the genesis of any transmission or reading encounter, that the "hyper-materiality" of what is going on is occluded, and aesthetic ideologies triggered. This organ or apparatus that intercedes in the production of sense, and socializes it, deletes traces of itself in producing, again and again, its product: one for whom face, mimesis, literalism, historicism and the "proletarianization of the senses" fill out the screen. That can be modified with different core referentials – different totems, different laws, bodies, "gods" and technologies – but it functions again and again in similar fashion. If I suggest, now, that this invisible organ, which recuperates reading and effaces trauma, resembles that done on the cinematic clips cut into place as "today's memory," it is only to say this would be hyper-compressed, a mini-apparatus so instantaneous and so automated that it takes almost no time and space.

 Borrowing a figure that appears in Faulkner with something similar in mind, I picture this machine of distillation as a moonshine still – that is, the apparatus with which hill people secretly brew their intoxicants, a big kettle with a corkscrew run-pipe called the "worm." The kettle resembles a stomach and the worm its exteriorized intestines, but this digestion process produces its own pharmakon. Today, in contrast to Faulkner's 1930s *stills* of the Prohibition Era, when alcohol was outlawed and legendary gangsters and locals ran the trade, spawning the great gangster era), the proper analogue might be a rural meth lab. But I retain the arcane image not just because it emerges in Faulkner. Moreover, at the right place and time, control of this apparatus would have meant control or dominance, the organization of memory in its proto-cinematic modes – still played in nationalistic propaganda media today – and the literary *pharmakon* itself.

In a largely ignored tale within Faulkner's mosaic of seven in *Go Down, Moses,* called "The Fire and the Hearth," we encounter an odd scam gone awry. The main character, Lucas Beauchamps, a Black tenant farmer, serves as a figure for Faulkner the author in trying to hide, and then get rid of, one of the components of his literary techniques. In *Go Down, Moses,* a certain going under is practiced against the Mosaic author and tradition of "the Book" – so capitalized in "The Bear" – that Faulkner's writing emerged from and against. Like the concluding poker game of the first tale, "Was," everything is gambled and on the table. In each of the work's segments, the logic of escape for the slave, or the Black, or the white is probed, and fails. Figures of the hunt parallel those of reading gone awry – the old plantation begins, to repeat, as an upside-down scene of paralyzed repeated rituals, including the repetitive play of the hunt itself, already without consequence, like a dog chasing a pet fox in the house. That is, rather than the antebellum South, Faulkner's "Was" opens defining a paralysis of contemporary and indeed modernist reading rituals. In "The Fire and the Hearth" Lucas's comic role in the dis-assemblage of Faulkner's writing apparatus is focused on the oddest of all of its components. The text will call this a "still," what distills intoxicants, what is the illegal maker of moonshine:

> First, in order to take care of George Wilkins once and for all, he had to hide his own still. And not only that, he had to do it singlehanded – dismantle it in the dark and transport it without help to some place far enough away and secret enough to escape the subsequent uproar and excitement and there conceal it.... The spot he sought was a slight overhang on one face of the mound; in a sense one side of his excavation was already dug for him, needing only to be enlarged a little, the earth working easily under the invisible pick, whispering easily and steadily to the invisible shovel until the orifice was deep enough for the worm and kettle to fit into it, when – and it was probably only a sigh but it sounded to him louder than an avalanche, as though the whole mound had stooped roaring down at him – the entire overhang sloughed. It drummed on the hollow kettle, covering it and the worm, and boiled about his feet and, as he leaped backward and tripped and fell, about his body too, hurling clods and dirt at him, striking him a final blow squarely in the face with something larger than a clod – a blow not vicious so much as merely heavy-handed, a sort of final admonitory pat from the spirit of darkness and solitude, *the old earth,* perhaps the

old ancestors themselves... – a fragment of an *earthenware vessel* which, intact, must have been as big as a chum and which even as he lifted it crumbled again and deposited in his palm, as though it had been handed to him, a single coin. (38, my italics)

Faulkner's "still" in the text above is to be buried in an earth, effaced. Yet that triggers "old earth" to be then exposed, saturated with time, and it causes to pop out yet another artifact or technical object – an earthen vessel with a gold coin. The clods disclose another machine. The imprinted coin not just money (yet) but gold, what triggers all mad pursuit and releases magical fables of hidden treasures, itself planted, a pure technic. Hiding the "still" – we are talking moonshine here – Lucas gets hit by another, homelier jug of sorts, what Wallace Stevens calls a "jar" whose poetic lineage runs elsewhere in *Go Down, Moses* to a citation of Keats's urn and another familial legacy. The latter is a container from which all coins had been replaced with worthless paper IOUs, as Isaac's bestowing Uncle Hubert's dissolute habits required stealing back from the future on empty credit ("The Bear"). Lucas is a Black tenant farmer renting now on the old plantation – the remnant of a hermeneutic plantation of rituals of hunts, of slave escapes and returns. The spell it is shadowed by goes back to its absent criminal patriarch, Old Carothers. Lucas's Blackness however cannot be ciphered in a face that precedes and moves behind or absorbs faces, antedates its original, defaces:

> the face which was not at all a replica even in caricature of his grandfather McCaslin's but which had heired and now reproduced with absolute and shocking fidelity the old ancestor's entire generation and thought – the face which, as old Isaac McCaslin had seen it that morning forty-five years ago, was a composite of a whole generation of fierce and undefeated young Confederate soldiers, embalmed and slightly mummified – and he thought with amazement and something very like horror: He's more like old Carothers than all the rest of us put together, including old Carothers. He is both heir and prototype simultaneously of all the geography and climate and biology which sired old Carothers and all the rest of us and our kind, myriad, countless, faceless, even nameless now except himself who fathered himself, intact and complete, contemptuous, as old Carothers must have been, of all blood black white yellow or red, including his own. (114)

About Lucas's "still," two things perhaps of relevance to an early twenty-first century scan of the *literary structure of climate chaos* – that is, really and in fact, the cause, if you like, of the disarticulation of the biosphere and climatic tectonics. You doubt me? Let me be precise by what I mean: I am not interested in how the new horizons, in which suddenly an "anthropocene" era is acknowledged (oh, my!) that is synonymous with ecocide and accelerating extinction events, or how we can now read these logics back into the archive from its inception (back, say, to what Bernard Stiegler calls "arche-cinema" – that initialization of perceptual orders, movement, mimesis, collective mnemonics, animation, the hunt, and the play of "light" on the walls of caves, sketches of animal prey thirty-odd thousand years ago). I am not interested, here, in a literature of climate change that is emerging with great difficulty in novels. (Unlike cinema, that is, where a certain climate change unconscious has been channeled but in a distinctly anaesthetizing way (someone always survives and renews the future): such anaesthetization plays into something that no one seems to note yet: the arrival of something like the twenty-first-century politics of (managed) extinction – and you were wondering why the ".00001 %" needs to now own all that stuff, from water to robotics, going forward?)

By saying literary cause of "climate change," I am aware I again invite ridicule – "You mean," let me say this for you, "this zone we have all moved beyond, literary stuff, the aesthetic, is actually where the impasse of ecocide resides or is engined?" (Lucas, here, would probably go silent, maybe even retreat behind a wall of clichéd Blackness: "Without changing the inflection of his voice and apparently without effort or even design Lucas became not Negro but nigger, not secret so much as impenetrable, not servile and not effacing, but enveloping himself in an *aura* of timeless and stupid impassivity almost like a smell" (58)) I will do the same, to a degree, since there is a Black optics in this writing (as writing) that I will draw on later. For now: even if the topos of "language" and climate change has been ignored, I will resist the long dossiers that need still to be assembled, many just practical in nature, to focus on something quite precise: the way in which, it would seem, a certain hermeneutical tick naturalized in Western discourse conforms to how hyper-industrial "man" is produced and reinforced as an ecocidal being – that certain ways in which meaning has been produced consolidate this, and even play to the forms of denialism staged or encountered today. To push a few buttons in advance, let me add to this a tincture of de Man and say, in a certain sense, that the guarantor of ecocidal acceleration, feeding into the elevation of corporate personhood itself, is the appearance and consolidation of aesthetic ideologies. What Lucas calls above his "still," which is also that of Faulkner, and encompasses the intoxication of literary writing and, at the same time, its reversal into a plantation order of hermeneutic rituals and consumption, is what produces this socialized form of the "we." The "still," in effect, must be set aside as addictogenic and machinal at the same time in *Go Down, Moses* – in which the gold coin will come to be sought by the "divining machine," a metal-detector apparatus lugged about as a coffinlike box, miming that meaning or revelation which the narrative reader might seek out. In his remarkable interrogation of his "rhizomatic" production, *Faulkner, Mississippi*, Édouard Glissant focuses on Faulkner's compulsory deferral of a secret that remains on the surface: "To point out and to hide a secret or a bit of knowledge (that is, to postpone its discovery): this is a great part of Faulkner's project and the motif around which his writing is organized." (6) When Faulkner trolls and exposes just this technique as a machinal technology to hide, conceal and lay tricks over finally, in which a "divining machine" becomes the trope of a hunt for meaning, for tired modes of reading, which might be lured by a planted coin whose discovery is staged, we witness the text expelling and trolling its own, past, "modernist" techniques and consumer culture (reading model).

There are, nonetheless, places where the question of language and climate chaos need be excavated, starting with the most obvious and leading through cognitive and neural settings: for one, the abyssal public nomenclature that surrounds these "climate" realities, this otherness without other(ness), a mix of dead metaphors and flat scientisms which corporate-sponsored denialists must have instituted as a poisoned gift; another: the streams of corporate algos and memory programs that disarticulate attention knowingly; and then, the "literary" origins of key cultural memes or ideologemes ("Nature" as a commodified misreading of Romanticism). And so on – all worthy, but I am after something else. What I do have in mind is more lethal and returns us to Lucas's "still."

Faulkner disassembles his own writing machine in *Go Down, Moses* – turns against the writer's own preceding production he questions as a "literary" game mastered within the era of what is called "the Book." The "Moses" of the title encompasses, or marks as a stutterer and forger, some first giver of inscriptions and "the Book" (as it's called in "The Bear"). It is compelled to put the former in its entirety in question, drifting into the black zone (Black voices and networked figures, blackness as what precedes "light" or the binary of white and black dependent on it – in the "old earth"). So it digs up, here, one of the machines of the literary plantation and the epoch of "the Book," or rather, it buries it, tries to efface or hide it, triggering a long narrative detour where a new, spiffier machine will be introduced directly to hunt gold coins, a metal detector called a "divining machine" and full of accessories: "an oblong metal box with a handle for carrying at each end, compact and solid, efficient and business-like and complex with knobs and dials." (79) This coffinlike "box" with "knobs and dials" corresponds to the hillbilly cameo of a certain AI module today.

IV

Now, I propose we update this algo. How is it that different utopist discourses, our best and most necessary to social justice, converge with their enemy netherdoubles along this general default: they could assign different ideals to a structuration of perception that defers, neutralizes and seals the ecocidal deal? And why would this question of "meaning" pursuit and production, induced by moonshine, frame twenty-first-century politics in one of its emergent forms – the politics of ecocide? Much as a season of polar vortices introduces a hiatus, or as Plato's earth in *The Stateman* that is arrested then reversed in timeflow, this excavation of allegorical apparatuses preceding "earth" – the worm, the kettle, the diving machine – moves us into a sequence of vortices that do not, as a circular model might, return us to a previously remembered state. These whorls spin off, yet may be tracked.

At this point in the review of the *literary* cause of climate chaos, I have focused on where Stiegler's account of the perpetual technogenesis and epiphylogenetics of "life," and his provision of an apparatus, arche-cinematics, that provides a mutating map for engaging it, is diverted from his "Simondonian" project by its premise, which stands outside phenomenality or descriptives. These are termed *hypomnemata*, forms of technics and inscription out of which a mnemonic or perceptual regime is generated. I have linked this use by Stiegler directly to what de Man references as inscriptions, thing-like and prephenomenal events and assemblages (think scratches in stone), mnemotechnic installations, against which cognition invents and diverts itself. The latter is not remarkable; it is how the "eye" functions routinely – selecting what it recognizes or seeks in a field of vision. One would need to put that regime in disarray to arrive at the type of seeing, no longer wired to the legacy of Plato's *eidein* (the verb by which he *forged* a unity of seeing and knowing – a movement to light). That is what de Man refers to as seeing "as poets do" in recasting the sublime with reference to the inversion of any human interiority machine (hermeneutic process) into the given accord of the visible as encountered – seeing from the nonhuman and non-living rather than the human thing, in a relational event of an eye "left to its own," an artefacted organ not selectively programmed (proprietized). This diversion into sight, the eye, visibility revisits not only perception as a reflexive product, something of direct relevance to a "society" that finds the biomorphic collapse and mutation in front of it invisible. It does not only ally with an arche-cinematics that, too, is in contact with prehistorial technogenesis but precedes the "Western" imposition of logos, alphabeticism, monotheistic-nihilism, and so on. One moves beyond (or can) the Eurocentrist focus that the term "anthropocene" crystallizes, with its Aristotelian *anthropos* as the core cartoon (again, male, Greek, endowed with "logic") – which then opens a counter-dossier to the Asian, pictographematic organizations of mnemo-politics and reference (a Sinocene, and so on?). This correlation of perception (in this case, what I "see") to assigned reference, and of coded referents to the authority of an artefacted "we," and each of these to the aesthetic ideologies that confirm these, seems to be at the core of the "anthropocene" fable. It is something that arche-cinematics tracks, regimes of reference that increasingly condense into capitalized reserves, "home" investments, claims of interiority (guarded by a phantom "we"), mimeticism, incorporation, eating, predatory induction into the same acceleration that consumes and vacuums up surfaces of information and earth today. So, there would be a question of the "politics" of disrupting a mnemo-sensorial regimes and archival automatisms: it is precisely this, say, that Benjamin does in his *Theses* by condensing the key enemy in the war for a possible future not to fascist armies but to the confirmation of an archival practice: historicism, a term that could be expanded to mimeticisms, empiricisms, agri-logisticism, the identificatory assumptions of face (cinematic spells) and the prescribed tracks of robo-engediture of a bad cinematicism ("proletarianized"). Or, more significantly, the reason why ecocide appears irreversible, prescribed by what may turn out to be a sort of hermeneutic accident, a fillip, an archaic organ of encryption that, like the irrational order of an English keyboard, arrived and opportunistically stuck, proved both efficient and helpful to shape and control a span of communities. Since it would be a pity to toss away an earth, from our perspective, based on something so stupid and accidental as old referential programming or an archaic apparatus for consolidating collective "perception," for encoding reference and for effacing the trauma of being the spawn of a techno-mnemonic bit of engineering – of being wedded to the marks on the cave wall.

To review this reverie: I excavated what de Man calls "literariness" as ciphered through Stiegler's *arche*-cinematics: an invocation and accounting for where mnemotechnics marks a space preceding the historical regime and interpretive violence projected and inverted by it, an address of the zone of inscription that represents a core formatting of a so-called present. This is a site, non-site, logic, moment or practice of defacement, as we saw with Lucas, and it returns as a key tool of the ecocidal thought and its streaming now. It is less cynical than cinecist in not identifying a "human" core, house, interior or position as its reference – it reads the latter's invention as a semantic fiction subscribing to certain hermeneutic automatisms. Telemarketing and propaganda bots had from its inception figured this out, as had cinematic practice and mainstream propaganda (Fox News's orchestrated repetitions spawning "fact," on up).

Defacement is practiced not as a violence against the cohesion of societal settings, but as the default logics that, in order to suppress the artifice of what he takes for his "own" world and property, triggers the ecocidal engines to overdrive. Our point of departure mimes the complaints that "current systems" are unsustainable (capitalism, "democracy," financialization, the petro-dollar, the "eco" in short), yet without replacement, and so doubled down on, rendered hyperbolic (quantitative easing, digital streaming), hypnotic, drugged and channeled (hacked, "proletarianized") – everything, that is, except the cognitive-grammatic product sought.

V

What, after all, is a *war over inscriptions* – the fuel rods of energy transference, organizing and channeling, to different ends, the inorganic. This is the supposed dilemma, since everything that contributes to this configuration in terms of mnemotechnic settings (or orchestrated "cultures of distraction," disruptions of "care" or "attention," "short-circuiting") is delegitimized by climate change and the advancing settings of ecocide that, now, define it, and would introduce a discreet politics of managed extinction – one which will be particularly alert to the role of populations, "failed states," disposables that traverse "state" borders and geographic mutations. The premise of defacement is no different than the other polar vortices tipped here in terms of "epochs": the stability of the last fifty, or then several hundred years in climactic terms, to say nothing of these ten thousand or so, was the exception and not the rule or norm, a parenthesis reverting to its volatile options of extreme climatic shifts and turn overs of life forms. It is not that we shift here from an identification of the eco with "nature" (organicism) to that of a technics we artifact as our interior or homeland or tribe (eco-technics) – it is that the artifice of this "eco" itself constitutes an economics of delayed extinction, as now evident. Moreover (and this is my contribution), that one can attribute all that to a certain archaic implant gone viral and awry in the default settings of the cin-anthropocene era – the era that organizes and generates its *anthropos* according to settings best suited to archaic tribal spells and politics (which, to be fair, can be said to return in the era of digital mafias, fragmenting nationalisms, corporate gang cultures, the gaming of all against all). It is not accidental, then, that one charter member in the moonshine still club, utopist progressivists, doesn't see their own structure mirrored in the disembodied supra-organism of the "corporation" that accelerates extinction scenarios because it does not account for human bodies – even if it mimes an apocalyptic anti-abortionist Christian ("Hobby Lobby"). Such a one could not "see" where its own blind, automatist distillation through the "worm," its weak messianism, prefigures the sheer technicism of the corporate "person," without face or body, that it recoils at in its bare, hyperbolic, monstrous, world ingesting form. Alternately, it cannot or does not read where even its seeing, or eye, is programmed similarly. It would not see that the trajectory of ecocide has no relevance to this automatism. If it turns out the hyper-computers to come do decide to exceed and terminate their messy, ravenous and unreliable forebears and petrie dish (hominids), the "corporation's" insertion as person at the heart of the state it powers will be deemed, in their history books, something like Germany crushing Brazil in the World Cup. This is where the control or desportment of inscriptions is warred over, as if the future of "life" on earth depended upon it – leaving aside the recuperative prospects of its future hybrid "we's." The logic of defacement implies that the only communality to be given space for, now, begins with the retirement of the robo-"we" projected to reset or redeem the human on human accord. Whither then?

The problematic of "literariness" surfaces once more before going under, and why not for good – but it disappeared a bit hastily not to be suspicious of what was being ushered out as merely aesthetic, at worst nihilistic (but in what revisionist sense?). Since what cannot even call itself a black enlightenment is unleashed now, and reclaims the banned zones covered by words like nihilism and cynicism, or literary, and diverts before relinquishing these roadbumps (what, after all, is a positive "nihilism" that aims against destructive or blind nihilist practices, which is what ecocide denotes?). The *Nethercene*, but outside of a shadow and anti-shadow (screen) set of polarities. One must be Blakean in this regard. Or Faulknerian, or Parmenidean – insofar as, de Man suggests, similar patterns repeat since there was "text" as such, similar modernisms staged, and the entirety may be thought as an encompassing (non)modernity, what had not yet been named "anthropocene" or indexed outright to irreversible ecocide (and extinction events). That would be the "era" of cin-animation as we eyeball it today. I will return to this eyeball, mentioned before, and suggest where that is accounted for by cine-photographesis – keeping in mind that the era of machinal cinema, which overlaps with that of accelerated hydro-carbon use (the "great acceleration"), not to mention techno-genocides, is an episode in arche-cinematics in which the former fully exteriorizes itself, materially, before the template dissolves into the redistribution of the digital transformation – in which mnemotechnics again mutates.

What de Man added to this tincture becomes manifest when the war shifts from the daylight of "transindividuation" projects to the *hypomnemata* themselves – the last line of defense against last man culture and the politics of extinction being architected (algo adapt, endorse subsistence, be disposable if needed for the survivors' pod to flourish – for the species sake, of course).

What does de Man's defacement of "descriptive" practices relay strategically? Does it relate to the appropriated quest to roam alternative and pharmacopic mode of "consciousness" today, to our favored or neuro-psychic entries into other zones (ayahuasca, edibles) – or Musk's video-game twitch to annihilate screen opponents (upgrading Trump's WWE model)? What derails the machinery of aesthetic ideologies from remultiplying, like "metaphysics" or retro-humanist turns (back to phenomenology, to new media as tool, to "we" extenders of otherness to the subaltern cat or stone), and the passive mimetic contracts to reference and identification it rewards, bribes with sugar rushes and exclusionary perks? How does an "irreversible" reading transgress the automated refold, domestication, assignation of reference? Or, the discreet mnemonic apparatus inserted into the production of the "we" – which any speaker can only cite and ascribe credit to in an act of perpetual accounting malfeasance. This recuperative apparatus, the moonshine still, is remarkably efficient, surviving revolutions and technological transitions with astonishing resilience – embedding itself in the site of transmission, canonizations, archival management.

Recall, individuals are burned or shot or jailed for their *interpretations* of phrases. Putin's Russia dismisses climate change as relevant, since it calculates future palm trees and resorts in Siberia and an Arctic pond at its disposal – for a while. No one seems to ever calculate or project beyond the end of this century, as if that were distant, as if getting there intact meant winning the race, nominally against a four-degree global heat rise yet in fact against an arbitrary sizzle of the screen altogether.

But it turns out that the apparatus, this naturalized implant, has predictably addicted its inorganic offspring – since this is where the "proletarianization of the senses" would be modified, executed in advance, or accelerated in response to conditions. Of course, "literature" would just be the epochal name for certain boundaries and canons to be shaped, sometimes reinforced or projected onto and used (or excluded) – and with good reason, since it is bound not only to unwinding the conundra posed by Aristotle recuperating tragedy from Plato's knowing if irrational dummy spit (of his closest semblant, the poet). This "literariness" is a function, a seance logic, the discreet site of transmission, cognitive artesinals, contract renewal to various mimetic regimes (or not), and what shares DNA not only with pure cinema but the effaced premise of "consciousness" as effect, that which has no organic existence yet drives, fashions, instigates and mutates "life" or animation. Contrary to being the only living thing with language, that human gift, it turns out the hominid branch – since a bunch of them are mixed in the delta of time – has what can also be the stupidest, weighed down with dead totems, infobytes, trained saluts, abstract "realisms" and retro-projected genealogies.

Now, I've kept the "still" in play so you will have a mimetic toy to sniff. I am addressing a you – but it is a composite you, one not quite singular, and including now the unborn that overshadow the "present," make it irritable, as if it were being asked to give something up. And yet this hybrid you without a we is sprouting all about. Worse, the more rewarded you are with the sugar rush, the moonshine hit of comfort, the more mature and expansive this organ becomes, this apparatus, this spectral muscle that squeezes out a packaged, grammaticized variation. It is not developed strongly or at all in "persons" close to agrarian subsistence economies or their urban survivalists. It informs currencies. It is, unwittingly, the secret sauce of the "short-circuiting" play which accelerates as tertiary retention (mis)takes its own previous technological version as its object (or prey – a negating and, hence, evolutionary move at inception). You may see this as the mode by which capture occurs, for which the hyper-computer or "corporation" has no need or even recognition: like watching small children play and talk to imaginary mates, one turns away. Right now, it appears bloated and drugged, insensate: the prison-house of artificed referents as, it seems, onscreen, sprayed as if with some chemical that makes everyone lean into literalizations of all sort to fill the blank spaces, doubling down. Again, unaware that what it is programmed to protect and defend, its very humanness, is and was a canard produced by automatisms. It was never that one couldn't get out of the "cave," even now, when it is a helmet of digital streaming and push-button memes. It was always a question of whether the torch would flicker enough to change or precede the template – eventually the public would get bored with the film, so myriad variations in it are introduced, ADHD injected into neural firings, various pills to compensate. And it will assure its creatures that if they ever leave this cushion, where their 401Ks are kept, they will not have a chance at all, disappear utterly, not have the tools for self-preservation or defense against the to-come – so they maybe reach a hand out of the cave mouth and drag in what isn't nailed down (pets, rocks), and feel they made their move. The problem, in the era of climate change (well, there is more than one), is that this same errant implant seems to fuel ecocide and its acceleration.

The moonshine still is behind Benjamin's angel, in petto, who is in the *Theses* a fraud, a failed prosopopeia in costume, the cause of the disasters "he" dissimulates before, looking back, the inert poster-boy of the Anthropocene instantly, before "he" gets to audition, is defaced and disarticulated, wings caught, by what the text calls, unread, yet over and again, a storm (indexed to "progress," coded as technological acceleration). This fraudster angel, who holds the reins of Marxist utopism and Hebraic theology in either hand, and takes both down with him, is the last creature to defend the weak messianism he seems, at first, to bookmark – until, it is clear to everyone but readers of the text, there is no messianism whatsoever, and "weak" means "rotted in advance." I say everyone, but modernist readers of Benjamin tend to repeat the two legacies that are trashed in the opening thesis (Marxist and Hebraist), heroicize the angel, make him into the model of the materialistic historiographer who, emerging from and with that text, bears a grapho-thaumaturgic promise to transform pasts, blast open temporal continuums, keep open futures whose gates are swinging shut. Bad luck that climate change appeared, and seems accelerated all but intentionally for certain ends. It is worth remarking here, then, that the entire conceit of intentionality itself would appear not with theological commentary, where a certain "God" would demand authority (a psychotic god or a modernist, nice-guy god, indifferent) so that his intentions were pretty important. That would be contrived, indexed, cryptically invented, politically decided without the philosophic contraption of "intention" being taken literally. "Intentionality," rather, could be seen as another piece of glue entering the era of stability following the Little Ice Age, platforming an Enlightenment only capable of emergence by leveraging that elite stability – fed into the sausage maker of the "still" as necessary. This moonlight still, out of which the artifice of the "we" is crafted, is not only a prototype for the "corporation" but a nihilistic crafting and channeling of reflexive interpretation, of intelligence.

This zone of inscription that returns, first, as an alternating or intervaled mark or sound, like those applied trauma of torture Naomi

Klein likens to "shock" capitalism – a stochastic disruption meant to break down consciousness and identity to plunder the memory. This is rendered kitsch in Hitchcock's *Spellbound*. In it, the discreet agon involves cinema's usurpation of the aura of psychoanalysis as anointed post-war manager of memory. It is what is itself tracked in the film as the series of lines or the seemingly vocal wails of the theramin's touch – circling a core amnesis in which there is no identity (the cinematic "psychosis" of Peck). But this insignia by which cinematics marks itself and knowingly puts itself in touch with its entirely nonhuman and machinal premise precedes any coalescence of word or face, is hypermaterial and without any mimetic corollary or even phenomenality, preletteral and without reference, and irrepressibly in motion, as if animated, as animation: fleeing antelopes and sabretooths, say.

VI

The trail leading from graphesis to and through climate change and ecocide is labyrinthine. Even the term "anthropocene" appears as if it wanted to post an inscription for yours truly that could be read, like the Nazca lines of Peru, by another eye from space. It obeys the crude rule by which whatever is inscribed may be acquired, "killed," inducted into nonorganic times and calculations of utility, memory organization and logistics, as a photograph does. The overlap between cinematic modernity and accelerated resource extraction lies just beyond the more obvious links to war. But the carbon link that binds ink to oil introduces a bond that is different from that between oil, as the giver of "light," and cinema – no oil, no cinema; no oil, "stored sunlight," no era of photography. No digital transition, no ecocide. As a techne of technics in the present fossil-fuel order, that of the energy wars that roil territorial claims (water is next), *oil* appears as the recycling of aeons of organic "life," a sort of cannibalism of "life" on itself for energy hits. This, with predictable dyspepsia and bloating (since what is the obesity epidemic but stored energy awry, fat cells mimicking oil's "stored sunlight"?).[46]

But a turn from letteration and grammar, to the order of inscriptions, to the hypogram and technogenesis of archivization, these may present as an ooze, an oil slick, a mode of transfer that, like oil, deflects light and shape.

The *hypomnemata* of Stiegler and the hypogram of de Man deliquesce into the hyper-material chains and technics tied, in petropathic leaps, to a viscous network and energetics. A certain circuit not only runs through a "proletarianization" of the senses, specifically visibility, or the "eye." But it does so by way of a technic of technics that is also, literally, situated at the inception of, the core of hyperindustrial "light" and the twilight of the Anthropocene. The Anthropocene is upon arrival in perma-twilight not only because the term has an irritating and quick half-life, appropriated already for geo-engineering propaganda. It operates as a de-extincted trope. It is in the "twilight" since the ecocidal scar it implies is not that to be deferred or put off except for being transposed to the timeless swarm of algos and info-streams on screens: "we" are (is?) the catastrophe (ask other life forms). Have you noticed a certain smell in the air today – not that of particulate matter or forest-fire haze, with the aftertaste of heavy metals, but of a literalism pandemic? It is subtle but unmistakable and skates up the spine of critical practitioners and public media alike, the latter with its practice of attention disturbance wired in.

The literal guarantees the momentary mirage of a real at issue, like hitting a button, and draws away from a catastrophic encroachment which shadows it. Oil is so literal it cannot be touched or held – it is toxic, waste, obscene, *excressential*, the pharmakon of energy transference at its most irreducible. It stands in as the black ops or circuit out of which the aesthetic is generated. And yet, while "responsible" for every cinematic projection, for machinal cinema itself (informing celluloid and graphematics), it seldom is marked or seldom marks itself within the frame or screen – as one might expect such a demiurgic god to reflexively do. If there were a network of vortices that returned from something like oil itself and if it preceded the *phenomenality* it gives rise to, one may attempt to access its positionalities (which look back on petrol-man, busy ingesting plastics at this point and thus, together with corn, its creature). One might attempt to activate a sort of petrolepathy that bears or transports us, vehicularly, into alternative planes and events, from industrial marking systems and cinematic capture, to resource wars (today, virtually all – ISIS, Ukraine, Nigeria, South China Seas, and so on), down to the tenderest urban entertainments, and so on. This occurs with an accelerated and addictive feeding (off itself). "Drill, baby, drill." And, of course, AI only ups the stakes exponentially with its ravenous data centers. Since the only bet in town is on whether some global governance might curtail oil (and coal) over the next decades, literally abandon or write off trillions of dollars of reserves before "alternative" energy sources (solar, new nuclear models, the Arctic, the moon...), we will witness a war with and over oil in its economic twilight, or not, as an additional theater to come.

To track oil's visibility is to ask how it appears, when it does, in the photographic image – how it is purveyed, too, when it erupts from its non-space, spills out into the surface, imperiling and ensconcing or reclaiming life forms, reported on (or suppressed) by media, as with the "Gulf BP disaster" of 2008. Or, rather, "spill." Like Negarastani's "blobjectivity," a center that assumes a depersonifying role over its human writings, or a globjectivity, the perspective of oil "itself," its leaking across membranes and categories of life and machination, addictive accelerations and ecocide, solicits the arche-cinematic zone that, too, precedes the artifice of "light." Any so-called black Enlightenment here gives way not to a blackness that is opposed to the rhetoric of "light," not only to viewing something like the Enlightenment template as ecocidally shaped, a black ops operation that shapes the "eye," but what precedes any installed division of black and light. As in Faulkner, blackness precedes any binarized referent (melancholy, death, "race"). Utterly exanthropic and yet informing "him" – no oil, no "Anthropocene." When Diogenes wandered about with an oil lamp in the day, looking for a man, an honest Anthropos (clearly, Plato, Aristotle and Alexander didn't make the cut), what did he know?

VII

Of course, Stiegler's positioning as post-deconstructive or the deconstruction of "deconstruction" resonates with Paul de Man, pre-deconstructive in one sense; does it imply that "deconstruction" for the era of climate extinction may never quite have taken place? One leaves out Derrida momentarily for a reason, though he, of course, informs and platforms these cancellations. Derrida's caution for deconstruction ("if it exists") is not rhetorical, at least insofar as that would be the deconstruction of something: of a complex, a history, a hermeneutic reflex called "metaphysics." The *après*-Derrida, stumbling about, was striving to produce a phantom Derrideanism commensurate with misplaced mourning rituals. It returns us to the sideshow in the American installation of the brand, wherein, as no one recalls, it seems, Paul de Man had initiated his interface with Derrida: that is, in deconstructing its claims, since any deconstructive reading would, minutes later, require to be in turn deconstructed, and the text – text as such – deconstructs itself in advance of any reading. This in effect did two things, as played out in the posthumous career of de Man's proper name, the burden it posed to the brand, his collective occlusion at the time (and, still, in quarters). It echoed in Derrida's subsequent treatment and occlusion of de Man.

The first implication was that this revocation of "deconstruction" would be buried in one of its suppressed genealogies, as a double-book accounting, and threat (the radioactive waste dropped in the sea, unapproachable or sleeping, wrote Derrida in *Biodegradables*....).[47]

The other returns to what de Man represented in this confabulation by turning the spotlight onto reading and inscriptions – what Stiegler would, in measure, name *hypomnemata*, oscillating between naming the particular technical apparatus of inscriptive mnemotechnics and the indelible, all but unapproachable inscriptions as such, a zone Derrida seances in his essay on *khora*, but in measure isolates there as an unfollowed-up-on-direction (the marketplace gets goggle-eyed). This immaterial "materiality of inscription" enfolds entire epistemes and interpretive regimes, antecedes figurative language and conceptual legacies, generates by evasion and concretization faux reserves and ecotechnic identities. *It* echoes Benjamin's inverse "materialistic historiography" conjuring a precessionary script – today, *coding*, a non-site from which perceptual and mnemo-technic "consciousness" is generated, deformed, hermeneutically captured, edited. This zone that de Man swerved toward, before and after "deconstruction," could be played out only by diving into prefigural complexes and returning with the debris of semi-articulated, de-anthropized structures calling the totality of hermeneutically designed interpretive strategies into relation to an absolute exteriority that "language" itself harbors (what Benjamin implied in calling language itself "nonhuman").

What de Man named "language" translates into mnemotechnics. It would turn out that de Man's silent occlusion within the final branding of deconstruction, or its appropriation by what Stiegler calls "little Derrideans" in the attempted corporatization of an *après*-Derrida, which has not turned out as presumed (less and less deconstruction), would take a position alongside two other zones that Derrida, in order to rhetorically stage "deconstruction" in his late work (with an eye, then, on his work after death entering the canon fully, surviving), structurally had to occlude – and they have a strange parallel.

From this deconstruction of itself that precedes and, in the ballet of time and branding, founds "deconstruction" as a proleptic specter (of itself), non-existent but having been, all along, the prefigural point of departure ("text" as such, inscription), several entanglements return to the present – wherein the Derridean corpus is re-read by the selecting algorithm we will call "twenty-first-century climate chaos," or what I will call simply "the worm. "

Nor do I mean what is nonetheless worth considering: that what would always be retro-projectively named "metaphysics," and congealed with an antecedent to go beyond, never existed as such.[48] (Derrida's ceaseless corrective, cutting off any scion's "post-deconstructive" claims, as in the case of Nancy's work on touch, would be consistent with a limit he guarded, an archival rim he systematically turned back from, in measure for his audience and in measure because these zones would, in essence, question the platforming of "deconstruction" – leaving us, rather, Derrida's vast intervention as a transitional platform, a transit station to pass through, inescapable but not to be rested within and not providing any teleological promise though rhetorically inviting or allowing such misprision in those too close, or imitative, e.g., the *après*-Derrideans presuming to guide or control this legacy and, despite their best efforts, predictably stalling out if not hurting it, e.g., by their inability to address the worm, climate chaos heard as the auto-dismantling of a biosphere, because Derrida would not have written on or addressed it, as if "it" exceeded the archival framing or rendered "deconstruction" itself a symptom rather than anti-venom of the so-called Anthropocene calamity.)

The late Derrida would in essence occlude, say, three zones – de Man and his "materiality of inscription," with its misanthropy; what we today call "climate change" or the "ecocidal acceleration," even in its then malleable rhetorics; and third is, cinema, itself, which he eschewed writing on, posting, so to speak, only interviews.

Derrida, who refused to be photographed early on, ended his trajectory performing in films and perpetually managing his brand politics and avatars, avoiding the cultural-waste dump litigated for "de Man" at the time and, to a degree, what openly broke with, or through, certain archival peripheries. Forgetting de Man – which is also to say, parochially, forgetting where deconstruction revoked itself at its launch – what Derrida systemically occluded would be, at the expense to his legacy, climate change, the automorphic unfolding of irreversible ecocides, triggered and passing to an entirely other temporal set of breaks and *Konvoluts* (Benjamin's term) than being "out of joint" approaches.

Would it turn out Derrida was, unalert, complicit? A lot of ink has been spent arguing otherwise, by supplementation. But perhaps this was less a critique than a gift, which went unresponded to due to the *apres*-Derrida dropping the ball.

Climate chaos and, again, cinema – which, in his lone late interview with *Cahiers du cinéma*, he makes it weirdly plain is at once to be dismissed (an "art of distraction") and yet the very template of active deconstruction and the very exemplar of spectrology as conjured. (It is not accidental that Stiegler, precisely responding to these occlusions, conjures as ur-pharmakon an arche-cinema as if "older" than arche-writing, accessing a multiplicity of trace types (promised and left unpursued in early Derrida). But then he turns all of this against the totalization and entropic closure of an Anthropocene and its rhetorical doom, which he articulates as a *Neganthropy* that is neg-entropic, an active diversion of the accelerating vortex which summons a long-absent component of any deconstruction to date, that is, the transference of energetics. Derrida would perhaps live too long and not long enough: too long not to have been aware and chosen not to engage this, making his occlusion distantly echo the dynamics of occlusion in diverse modes of "climate" denialisms, however unwittingly, and too soon to have so re-oriented, at the cost of the "deconstruction" he had assembled across all his work. What did not emerge was any attack on the marshaling, diversionary rhetorics of the climate era and the Anthropocene in particular. The absence of the latter, and the frozen pieties and sterility of the *après*-Derrida, left "Derridean" archivists presuming to manage the franchise. They dropped the ball, distracted from the collective entropy of world-extincting forces crossing tipping points, after which the biosphere would alter, self-doom, become a sort of cosmic event (as far as sentient "life" in this parochial galaxy is concerned) – let's leave aside the status of European anthropism, terror, the death penalty and the mainline territories Derrida patrolled and trolled: religion, politics, ethics, whose academic seductions he would calculate as anchor babies for his afterlife. Nothing has wounded Derrida's legacy or perceived relevance more than his occlusion of engaging climate chaos. This, because it moved the limit he had set as a sort of contract with a future reader outside of the Anthropist backboard he deconstructed. The blind here raises the question of its participation in the techno-ecocidic trajectory (hence, the requirement to deconstruct "deconstruction," pace de Man).

This lends a powerful subtext to Derrida's final interview – what would be first titled "I Am at War with Myself" and which his English hagiographers deemed perhaps too disruptive and retitled, in a manner that lends itself to romantic affect, "Learning to Live – Finally." Finally, I have arrived, perhaps, just before death – or, well, I am at war, as I look back, with what I have done, with a turn I had calculated and been trapped by rhetorically, to the extent he had been deferring this other engagement, which he would have been aware of, and which awareness lends a certain bad faith to the strategy, on which the brand "deconstruction" would depend as a promise, of occluding climate change and cinema (and, by affiliation, de Man). That is, an affiliation indexed to the latter's turn against the discourse of trope altogether, his seancing materialities of inscription out of which perceptual programming, ideologemes and perpetual mis=literalizations are generated (stupidification), his de-anthropist vanishing points. It turns out Derrida would not, in a specific and traceable way, be deconstructive enough on this one turn, point or limit: and hence the phrase, "I Am at War with Myself," if "myself" represented a certain Derridean edifice, encyclopedic, "deconstruction," which "I" (Derrida at that moment) bifurcate from, separate from, identifying at that point with what

Stiegler conjures as a neganthropocene space or project – what channels, invisibly, the logics of spectrologistics turned against the totalization of the High Trumpocene. It is the same "last" or "final" interview in which he reflects grimly on there being almost no one who reads him properly (pace the very category of "Derridean") and that his project would disappear with him, as if either his living charisma were required to keep the project in play, counter to the very premise of arche-writing, or he had zero faith in the missionary archivists on standby to do other than relay to a reader *to come*. Derrida at war with Derridean deconstruction, as assembled, as it were: finally?

Which is which? Why would "Derrida" have been at war with, well, *moi-même*: that is, what chosen direction or rhetorical structure did he occlude that cannot be occluded any more without constraint, swerve, calculation?

Momentarily, since it depends now on an arriving generation to update their boomer professorial carriers, who have been swindled by their future-despoiling fidelities, complicity and blinds, in part because Derrida had failed to be deconstructive enough. They, these arrivers, would side with the Derrida at war with the folio he had just closed, aware it was closing, that the lights would turn out shortly. They would side, that is, not simply with the completion of de Man's revocation of "deconstruction" as a narrative episode or promissory, a revocation of the anthropist screen before an irreversibility it structured itself precisely against – and which hermeneutic reflex, sometimes called "mimeticism," would be inextricable from the techno-ecocidal acceleration. They would move to the space, perhaps finding other names, that Stiegler has opened as a spectrographic activism and digital war zones of a channeled Derridean innovation and a spectrology of the real that would project the twenty-first-century tools of the war which overtook that of knowledge systems and power structures whose debris would recombine, in the engineered retro-imaginaries of tribal ressentiments and their infantilized regression of the telesphere to iterative memes and its own "alternative facts."

To cite just one damage which Derrida's delay spawned, and the narrow attentions of the *après*-Derrida to look up from regurgitating Derrida's texts. They might have turned, at once, to deconstruct the rhetoric of the public spheres and the new situations of "language" following the digital transcriptions and mnemotechnic accelerations.[49] They might long ago have pulled the curtain on the tropological distractions of spell-like terms engineered for failure – "climate change," "global warming," inert faux scientisms that cool the public cortex, as opposed, say, to the ecocidal acceleration. But what again, might "I" have been at war with, in bifurcating from *moi-même*, from the

"Derrida" who had arrived, finally, at the cusp of death at... having learned to "live" finally (organ, please) or bifurcating against the entire edifice built upon a managed occlusion that, in the end, contaminated and drained other interventions if not "deconstruction" as a fable ("if it exists"). I remind of the obvious desideratum today: what of Derrida, selectively, is of import from the perspective of the "present" or non-present of a post-tipping-point Anthropocene – or, more simply, from the perspective of climate chaos itself, which opens onto a problematic not only of the inorganic and biotechnics but the data-ization of trace – and, most specifically, a turn against the totalization of horizons to which darts are thrown today, lamely and inauspiciously, against some genetically mutant capitalism (rogue, zombie, extinction economics, apocalyptic capitalism and its cartoon variants)? It is precisely against this totalization of tropes, now become algo-memes, that de Man withdrew to a facticity of inscription, what Stiegler's arche-cinematics actively or Derrida's *khora* active-passively puts into play, opening to erasure and dis-installation as the premise for epistemic recasting. AI, "who" now speaks to us in adaptable postures, yet is faceless, bodiless, without perceptibility and immune to duration, would recognize all of this.

One might draw up a list, entirely improvised, of what this other Derrida might war with in the one then bifurcated against, the "Derrida" of the movies, occupying the pop frame – a list not to be pursued now. Really, Jacques (*moi-même*) – deconstruction "is" justice? We are to hear about the death penalty, and that the world ends with each death but not the karaoke of climate extinction? Of "nuclear criticism" playing the Cold War fable of instantaneous disappearance but not biospheric mutation – precisely in the window tipping points would pass? Why do we vanquish the trope of Levinas's "other" early only to revive and puppet it in late twentieth-century currency as a *fait*? Really, hospitality, and *absolute hospitality*? We can tarry over the human agency of terror, but not the ideological trap of an Anthropocene without face? We can speculate on a politics of memory without assaulting tele-inscriptions openly, or take seriously the brief neo-con triumphalism and leave off of your ten plagues any reference to biospheric collapse in one of its memes? This repeated short-cut of the "X without X" (materiality without matter, messianism without the messianic, etc.) testified to the shortcut of the app. And so on.

One of the high points of *après*-Derridean comedy would be the first book after Derrida's death to gain and merit attention by the family franchise, Martin Hägglund's *Derrida's Radical Atheism* (2008): savor this, that the first thing Derrideans would need reminding is that Derrida was not a theist – and specifically reminding by name the Derridean archivist. Then, to top it off, Hägglund resolves that for him the real core of late Derrida is the ethical – which is a bit of an odd flash, since if melted down what emerges is a MacGuffin: choose, and try for the least bad option. Er, no... So there, already long after Derrida's death, one had to remind Derrideans that, no, he wasn't a theist? Another high point would be Geoff Bennington's cry of misery at finding old allies already lazy and inept in repacking Derrida for recycling, in the case of Gayatri Spivak and Judith Butler's marquee name updates to the re-issue of *Of Grammatology* for Hopkins – a cry that, nonetheless, had long resonated within a closed circuit of pledged Derrideans unable to move beyond repeating and looking for the Derrida of their earlier encounters and seminars, as if time had stopped rather than radically accelerated. Hamlet's "time is out of joint" seems now quaint as we precipitate into irreversibility, post-tipping-points vortices and multiplaned, temporal clusterfucks.

Why did Derrida, who virtually conjured the interface between writing itself and the anthropic trajectory, not complete the trajectory and address the inky orders of carbon and blackness of oil? Or did I, J.D. in the last interview, splitting from and at war with a strategy of readerly calculation, too reliant that he had already established all possible disruptions that the reader would import back into a Trojan-Horse strategy gone awry. Derrida shifts, in death, from the performative pathos of Hamlet to the position of the ghost, and an impersonal "visor effect."

In this sense, one might gesture to an arriving generation of Derridean readers – like the Parkland teens calling BS on their kleptocratic boomer despoilers, the most infantile hominid generation to date – that they might initiate this other reading, too late, that *Je*, J.D., targeted, That is, one which would selectively transpose or re-read Derrida, today, entirely from the lens of what climate chaos demands in its radical (non)present to reassess not "the political," or "the ethical," but what Bernard Stiegler calls "disarming and re-arming" of terms, context, agency before twenty-first century *totalizations*: Derrida would be right to be at war with the late Derrida, who, in actual effect, would *function* inadvertently as a climate denier and defer, in effect and by extension, precisely the powers of cognitive rhetorical deconstruction being turned on the Anthropocene stupor rather that retreat to memorializations. It raises again the question whether "deconstruction" as received was a disruption or an embellishment and participant in the late Anthropocene enclosure.

VIII

Stiegler makes a move here – to reclaim from any neurological transcription (for instance, Catherine Malabou's descriptive claims deploying the brain), the spectral locus of the "reading and writing brain," the locus of "self" as reading and the suppression not only of technics – by which Stiegler inserts "tertiary memory," or the facticity of expanded mnemotechnics as relentless exteriorization (of, one must add, what was never interiorized to begin with).[50] It is here that de Man (at the suspended beginning) and Stiegler (determinedly *apres*-Derrida otherwise), minimally converge. Yet it is also here that de Man further complicated the cluster-transition Stiegler identifies as a ghost ensemble in the gaps and compulsive error between inscription and act, between mnemotechnic and voiced speech, between relapse (the positing of an "I") and contamination, between inscription and phenomalization, perceptual-cognitive programmings rooted in reactive hermeneutic algos (historicism, mimeticism, referentialism). And here, at the far side of a vaporized "deconstructive" narrative, there is a fall and a split, as the *Trumpocene* disorder cancels the neat Anthropocene imaginary, and as mass extinctions expose the "wager" of this entire gameboard – hence, Derrida's critical lapse at just this juncture resonated. Imagine, if you will, a horde of trained, critical troubadors descending like a swarm on the long-engineered-for-failure discourse of climate change, the Anthropocene, and so on, in the marketplace in the mid-2020s – rather than building its swerve into identity politics' rhetorical dead ends, tributaries of the Trumposphere in default, because "Jacques" had pointed there and supplied a lead? But no: we were to get rehashes and hagiographies for this milling army, and wars for "family" capital, archival propriety, to the point when a stalwart among them offers a cri de coeur projected onto slovenly once-allies: are we embarrassing ourselves yet? That aside, that little window of mutual betrayals between a certain J.D. who, at the last moment, turns against the familiar in its entirety, is at war with it, as if standing apart from the entire frame and strategic choices, due to what they had, uniquely, begun to occlude as the cost for the Trojan Horse's canonization he required. Thus, Stiegler's commits to a positive organology of inscriptions, an artful deployment of the pharmacological prowess, the shaman powers, to ameliorate the dilemma of the *we*, the tribe, the anchor of the individuated psyche or transmissional soul, and to navigate what is called a future, but is perhaps a delay or deferment of entropic acceleration, requiring an alternate time-space it envisions. Yet Stiegler is occupied, at this point, with undoing the blockage and burden of a post-structuralist legacy complicit in an accelerated and totalized horizon of divestiture and stupefaction, "general proletarianization," and with mnemotechnics overshooting the digital transmutation, and with the "social" cloud of transindividuating and extincting spectro-political forces, epochs, breaks in the prison.

The split between Stiegler and de Man at the polar vortex before and after "deconstruction" is that between arche-cinematicism, Stiegler's trans-epochal reading apparatus that would be "older than arche-writing" technically and, aware of its double-edged pharmacological powers, reopening Plato's pharmacy in an era of industrialized synthetics swarmed with generics, knockoffs and cartel smack – and what one can call a sort of "arche-cynecism," if I can restore the other Diogenes to this moment, if he ever left, that is, where the tether to a "positive organology" is let slip for what might be called the "endless drift of a negative organology of inscription," for which the effort of restoring future, protention, promise (categories complicit in the ecocidal algos long at work) is set aside. Stiegler ignored climate change and biomutation up to a point, and this is why the subsequent move against the Anthropocene imaginary as a dead-end trap, in adding this turn, brings the march of Stiegler's elaboration to one core of what had been brewing beneath an "organological" writing template, as action and act, which is to say, that after which all else is compelled to reorder itself. But one might say: cynicism – even modified by Stiegler's radical restatement of cinematics as an epoch concluded with the digital but encompassing that of "the book" all along, and the letter – is irresponsible, like de Man, whom one dutiful professor blamed Trump on in a *Chronicle of Higher Education* column. The turn, then, to *arche-cinematicism* as to a site of pre-inscription – what Stiegler projects as a positive *organology* – pivots against the "Anthropocene," in a mutating hall of mirrors. He looks to reconjure protention, futurity, a Negentropic diversion and pharmacographic maturity; one might find the de Manian version a sort of negative organology, though it would not claim the latter's apparatus so much as sink, further, to spelunk at what has been overlooked in the order of inscriptions: what I term "arche-cynecism," as the option to roam the distinct technoscapes evolving and extincting, foaming, cast over and fueling irreversible accelerations. The arche-cynecist is skeptical of referential spells, and suspects that the current techno-ecocide trajectory has been as much embedded in the hermeneutic apparatuses' running on autopilot, the relapse, out of which the mnemotechnic retro-artifice of a spoken "I" emerges, embedded and concealed in others' words. And the arche-cynecist surveys with night goggles what lies outside the lit area of "Anthropocene Talk," its magic bubble of data-fied neon, aware that it must grow comfortable with irreversibilities without mourning.

Derrida would indeed be an "event" – as the current clime of abandonment does not put in question – but as perpetually transitional.. One, moreover, that generously contained a strategic lacuna, the biomorphic tempophagy behind the industrialization of "metaphysics" as a control strategy, which required of the *après*-Derrida yet another supplement to the supplement. One might call this "the little neganthropocene," a sort of infinite prelude to the not-to-come of the Neganthropocene and Negentropocene which Stiegler proposes, projects, lays inscriptions for, experiences as in place, yet would be impossible (not just "improbable"). This, at a moment when the membrane between existents and non-existents has frayed.

I am at war with myself? Not him, *Derrida*, this time, but "I," the entire interiority imaginary of subject-entities, on earth, with earth, with itself – and, apparently, every other species. I am at war with myself, me too, as I reject my own previous reading, that mourning but paralyzed Derrideans played a role in this closure, since Derrida foresaw and discounted that, and turned upon *moi-même* as if to index the stepping out of his frame that could only be done as the dissolution of a "deconstructive" claim, narrative, promise or brand.

III. DIOGENES'S LAMP – or, CINEMACIDE

Tempophagy

Hypothesis: The totalization and fragmentation of the digital-psychic escarpment, which Bernard Stiegler uniquely turned to confront, yields a "politics of managed extinctions" by default once tipping points have passed. And that Stiegler's conjuring of an arche-cinematics as a reading before "reading" mode of fusing a long paralyzed deconstructive vector to the climate doomloop acceleration, without promise.

8 Arche-Cinema and the Politics of Managed Extinction

Few progressives have turned around to face the future; and one can see why, for the progressive who turns around can no longer be a progressive. In the Anthropocene, in addition to the past we seek to escape, now we have a future we want to avoid; so we are squeezed from both ends.... The most striking fact about the human response to climate change is the determination not to reflect, to carry on blindly as if nothing is happening.
—Clive Hamilton

Many scientists concede that... we are on the path toward a 4C rise as early as mid-century, with catastrophic consequences.... Worse, [they] now project that we are actually on course to reach global temperatures of up to 8C within 90 years.
—Nafeez Mosaddeq Ahmed

The defunctionalizations and refunctionalizations that determine the rhythm of the organological genealogy of the sensible and of what lies coiled up there – the intellect and the unity of its reasons, its motivations – have specific folds that create ruptures that are called epochs, and that accentuate more and more vividly as time moves on the faultlines, the disadjustments, the incomprehensions, the crises, and critiques.
—Bernard Stiegler

I "Today": The Movie?

Somewhere in movie heaven, it is written: that shall only have reality which has been put into film.

I ask you to consider, for a moment, a Hollywood script proposal: we'll call it "20 – : The Movie."[51] It is a bit clichéd, and I don't know what to make of it (you decide). It starts with a premise: a hyperindustrial civilization receives numerous alarming reports that it has passed tipping points of toxic global warming – and that it has now entered a prolonged period of mass extinction events (including this species's own). Yet citizens appear unmoved, distracted or in open denial, as in a spell.

To continue. Extreme weather events escalate (megadroughts, polar vortex, etc.), resource wars advance, nominal democracies rip across the globe, utopic progressivism is in disarray. Even our utopist critics are reduced to amazement at this spell – if we define utopist criticism, in Jameson's revision, not as those who believe in the arrival of a redemptive utopia but those, merely, who struggle for social justice and progressivism (a downgrade). Here is Henry Giroux on this spell:

> The organized culture of forgetting, with its immense disimagination machines, has ushered in a permanent revolution marked by a massive project of distributing wealth upward, the militarization of the entire social order and an ongoing depoliticization of agency and politics itself.

Revolution has been permanently inverted. This description references the effects of media, telemarketing, television, cinema particularly ("immense disimagination machines"). It recalls what Bernard Stiegler calls the "proletarianization of the senses" themselves.[52]

It may seem odd to bring what's left of the American left in contact with Bernard Stiegler's writing of a "general organology." The former seeks orientation in a muted American political-scape, often appealing still to a democracy that might be taken back. The latter writes from a post-democratic, hyper-industrial escarpment which requires a thought simultaneously in contact with pre-historial technics as well as the capture of digital culture. Yet when Henry Giroux describes the dismemberment of American sentience and the "politics of disposability," there are echoes and rhymes with what Stiegler diagnoses when he depicts the miseries of a loss of "spirit" (of capitalism), the "proletarianization of the senses" (preferable, in English, to "sensibility"), or regimes of "dis-individuation." Stiegler uses Gilbert Simondon to mobilize against "dis-individuation" and the theft of "knowledge of life" by the outsourcing of memory. He thus draws any contemporary malaise not into a narrative of social struggle within a co-opted democracy but a politics of mnemotechnics and its epochal digital mutation, as well as the capture of perception – or the senses. If, in Henry Giroux's case, this points to "a new form of hybrid global financial authoritarianism," for Stiegler, even this is wired to forms of "short-circuiting" that enforce the psycho-technic disruption of attention and care. If, in the first, a "politics" seeks its own image somewhere, in the latter, that has migrated now into the neural paths, sensory programs, grammaticization.

This configuration is convex. It marks both Stiegler's reprioritization of aesthesis, the technogenesis of perceptual "consciousness" and its ill, contamination, default, and prescribed vulnerability to capture – and not just preemptively by the NSA or databots. If the "organs" of the senses – say, the eye – are constituted differently in different technical epochs, they have always been subject to being systemically hacked by the same back portal. For the individual memory that has been outsourced is not returned to the dependent, with a consequent theft of "knowledge" and specifically knowledge of how to live (savoir vivre). The "proletarianization of the senses" accompanies a "short-circuiting" effect, which today is allied for Stiegler to the collapse of care and attention, mass "dis-individuation" and the accelerations of mafia cultures (and, if I may, ecocide). In the case of arche-cinema, that ur-modus of "consciousness" initiating movement and "the organization of the inorganic" as programs of perception which antedate writing by tens of millennia, yet in which cave we still dwell, he elaborates what occurs: "it is the primary and secondary identification processes, which constitute the condition of formation of the psychic apparatus, and therefore the condition of production of libidinal energy, that are effectively short-circuited." Thus, while one can adjust the import in contemporary terms to absorb Giroux's metaphorical wail (and much else), it is not just the effect in question of hyperindustrial speed, mediacratic saturations, telemarketing bots, last-man culture, and so on. This "short-circuiting" could have arrived with the advent of tertiary retention, that is, at the scene of initialization, with marks and shadows on a cave wall. It could occur as instantly as tertiary memory remarks itself – a negative condition for the evolution of technical objects which mark, and negate, their last iteration, like the end running of high-frequency trading.(The association of technics with tool-like supplements – the inorganic – omits "its" default and spectral emergence with theft – perpetual as well as pre-originary.) The shadow realm of pre-originary technics occurs as theft, and theft of itself in turn – a theft that precedes that of "fire," and fronts the latter as a misleading emblem of itself. In a way, when tertiary retention hits the mirror stage and takes a selfie, short-circuiting is triggered. (The term "anthropocene" operates in this way, which accounts for its surge of popularity – it is a short-circuiting selfie of a sort, an impossible-to-close circle.) If the "proletarianization of the senses" leads to the inability to see what is before the eye itself, of which "climate change" is a prime example, it leads us back to where, in Stiegler, the apparatus is installed. It leads back to cinema, or arche-cinema.

Two discreet events inform the above mentioned script. It would be remembered looking back from the future as a watershed date. The first event: the dust finally settled following the "2008 financial crisis" to disclose a massive wealth transference engineered globally – instantly creating a bifurcating class system, or new "proletariat" or "pre-cariat." The financial commentator Catherine Austin Fitts speaks of this as a "breakaway civilization," the "super elite" or fabled ".0001 %." At the same time, a second event quietly occurs. Western countries discreetly back off their carbon-cutting commitments, purportedly due to economic pressures – implicitly acknowledging the irreversibility of catastrophic global warming and coming mass extinction events. The rhetoric will have discreetly changed from mitigation or even sustainability to something else: the new meme is that we will have to adapt and, moreover, that geo-engineering will aid everyone (a prospect bringing immense corporate profit).

These two markers – massively engineered wealth transference; acknowledgement that irreversible tipping points have passed – link up in this script. That is, they appear coordinated even as "climate change denialism" itself rises in the Anglo (or Murdoch news) nations. We can now see why, and it is rather bad Hollywood, but perhaps that is the point. It should be noted that this new "proletariat" is no longer to be seen as oppressed labor. Moving into an era of managerial robotics, there may be no great need for "labor," hence less employment. Moreover, as this massive wealth shift consolidates, in which eighty-six individuals own the wealth equivalent of the bottom global 50%, resources are increasingly being sequestered by the few and won't be returning to any commons. What the cinematically spellbound populace is unaware of, since it is their "senses" that have been proletarianized, is that they do not see what is, literally, before their eyes daily.

It is obvious, here, that despite the streams of corporate media and climate change *occlusion* (more structural than "denial"), *they* knew, and planned accordingly. One could perhaps now re-read what theater like *The Paris Accords*, and similar kabuki theater signaled. That is when all the world leaders got together on this and walked away, squabbling. Looking back, it was not just squabbling. It would have been an implicit decision. These "elites" could not make a radical turn in carbon reductions without losing their own political regimes. But instead of pulling back, they would unsurprisingly accelerate all carbon consumption. The decision had simply been other than expected: if one could not preserve a future with resources for the many, a few would consolidate them and form the survivor class. They would be aided by new hyper-technologies and genetic engineering, which would be privatized. "They," a certain corporate and financial "super elite" (not my words), would anticipate what the CIA report called "population culling" or a mid-century die-off, after all. Clearly, there would be no need to inform the population, since there was nothing they could do. In the film script, the new "proletariat" is no longer Marx's dialectical and revolutionary force: it is no longer defined as labor, since employment will be scarce anyway (robotics will reduce labor needs by 40%). This new proletariat, then, is marked as a disposable population. It could be harvested for metadata and wealth extraction (or, occasionally, body organs).

If you want a better Hollywood script out of this – have we seen this movie? – one might add for flavor that this "breakaway civilization" or economy (employment down, markets up) amounts not only to a self-chosen survivor class. Inevitably, at a time of exponential advance in genetic engineering and nanoscience, it implies a financially engineered species split (as the phantom of a "singularity" echoes).[53]

II The Politics of Managed Extinction

> There is arche-cinema to the extent that for any noetic act –
> for example, in an act of perception – consciousness projects
> its object.
> —Bernard Stiegler

For Stiegler, there are two poles of film practice, and, between them, they negotiate a sort of war – the stereotype (which includes Hollywood, in which cinema confirms familiar categories of identification and reference) and the trauma-type (which puts cinema itself into question materially, defacing the former). The first accords with his update of Adorno's "culture industry" into the "consciousness industry" of today, in which image programs, telemarketing and the implantation of memory are practiced. While this polarity seems at first slight (a binary for cinema?) the pair posit, as said, polarities between which "negotiation" occurs. The trenchancy of this divide reflects, however, the forms that arche-cinema – the cinematic template that, for Stiegler, antecedes writing by millennia and is initialized in cave paintings (more on this below), and not only platforms "consciousness" (whose cinematic artefaction occurs hourly), but melds with the lightless zone of the dream or hallucination. In rhetorical terms, it precedes the coalescence of face – in what is still the projected cast of marks and shadow. It precedes *prosopopeia*, affect, movement in one or another individuated regimes or grammaticized epochs. Thus, what is called a "stereotype correlate" to the "proletarianization of the senses," citing and reiterating "commonplaces" for recognition (and communal engineering). The other polarity, trauma-type, would deface the stereotype or short-circuiting. Here, cinema marks itself as a lethal, spellbinding apparatus at the service of homogenizing powers and the antithesis, trauma-type, which takes the Dionysian position over the Apollonian form of its other. The word "trauma" refers to and is in contact with the encompassing trauma and default of "epiphylogenesis." When Stiegler says that cinema is "life," he does not only mean the schism by which our consciousness of being alive, like that of "perception," occurs through what is not living – being entirely technogenetic, in default. Moreover, the machinal cinematic culture of the hydrocarbon and industrial eras was just an episode, an exteriorization of agency in a 36,000-year parenthesis – one ending with the digital transformation which dissolves the anagogic or celluloid inscriptions placed before the electric bulb into digital algorithms and interfaces. The trauma-type knows all this, marks and puts itself (and the hominid in the frame produced with and by it) into question, battles with the stereotype ("Hollywood"), and is acutely aware of its technicity and destructive prowess. How, nonetheless, deferring elaboration, do these "types" resonate when applied, today, to the cinema of "climate change" and ecocide – since today, despite all the denialism and rationalization and disinformation, it would be safe to say that everyone knows, that every organism knows and is mutating (again, just look at the bodies in any airport lounge), and since what is suppressed in turn creates an "unconscious"; even in public space, one would note that the entertainment-industrial complex (McKenzie Wark's term) needs to manage "climate" anxiety, "climate" info-debris, or now what is called "climate" brain, as you like. Since we are considering whether the script proposed above should angle more to stereotype (Hollywood remuneration!) or trauma-type, the question of applying Stiegler, even in this kitsch register, returns to the politics of extinction that the proposed script, itself, made its motif, theme or pitch.

As regards a cinema *of* the Anthropocene, or of the era of climate disaster, the first category seems today marked – no matter how refined the product – by several traits: the disaster film can be exploited, it is apocalyptic or post-apocalyptic (if we understand that as a residue of *Christological* thought, where something happens in a flash and it implies revelation or disclosure or even judgment). More importantly: someone survives to renew the world and identify with. The problem will be that there is nothing apocalyptic about catastrophic climate change – it is slow or sudden, reverses polarities, subtracting as it goes, in itself meaningless (it has seen this movie before, many times). It has no survivor or renewal, since it wipes out the condition of life as we knew it.

The self-trolling film *2012* was a Hollywood blockbuster condensing climate catastrophe to a single day – from which escape would only be had on a Chinese-built Noah's ark at the price of a billion dollars per ticket. In contrast, we might speak of "202X: The Un-Movie" because it is rendered irreal by its cinematic tone. Thus, the recent film *Noah* seems to invert a bit. In it, the Creator apparently delights in flushing mankind away altogether: that is, in its hedged way, cinema turns against us – but has an out. Of course, he does get to survive and, of course, breed again and renew the species (who mess up again, of course). And this may be the problem with climate-change films, Hollywood cinema in the Anthropocene when it channels this immense dread of being, already, in the afterlife, beyond tipping points. First of all, it favors disaster effects, even if these fail to shock, and it favors apocalypse to do so – that is, the Christological model of the sudden flash of revelation in disappearance (like nuclear blasts). Then there can be a post-apocalypse, where someone survives and revives. Someone is there to tell the story and repeople, once again, *the* future.

If the film *2012* ends with oligarchs buying tickets for a billion dollars each on a Chinese-engineered ark (only the Chinese could build it, but I assume they rushed to Macau to spend the windfall quickly), parodying the hidden logic of "20–: The Un-Movie," the film *Noah* reverses this with his real *Ark*. Making the Biblical Noah into a comic-book hero (appropriately), the Christian audience misunderstand the film as theirs: it is not, God is not mentioned (he's called "the Creator"), and he is nasty as well as absent. He's sick of humans, who have wasted the earth. He wants to drown and extinct them. Even Noah gets into this – but is diverted by his kids, who want get back into the game. That is, start another cycle of life, which would bring us to the present, when it all should be extinct again (as in the film). Nonetheless, there is renewal. Even in the film of McCarthy's *The Road*, a weak film in which a nameless father and son roam a dead earth after a cataclysm that is unnamed, the boy is taken in by a family with a young girl (there is a future, maybe, depending). Even the post-apocalyptic get one last next chance. One could add to this list diverse variations, including the Pixar or Disney animation products, like the Ice Age franchise (in which funny Mammoths make jokes about their own extinction) and *WALL-E*, in which a garbage robot on a future wasteland Earth brings life (and humans) back to the poisoned planet. Even *Avatar* falls into this camp, with its cynical deployment of a native American romanticism and Gaia-esque setting – since some argue it was romantic organicism itself that propelled us into the hyper-industrial era of use and extraction (of what was called then "nature"). The "consciousness industry" is working overtime on this. It suggests another open secret today, what can be called a "climate-change unconscious" – if an "unconscious" is created by occlusion or suppression, by decreed invisibility. Ignore all the "climate-debate" stuff, the denialist programming, and so on: everyone knows, because every living organism "knows" (micro-organisms, amphibians, polar bears, of course), shifting habitats, extincting, chocking on toxins. It is not that our favorite hominid is not mutating, as can be confirmed at any airport lounge in America, just that he practices dissociation and cognitive aversion.

But, of course, extinction has no remainder, there would be no renewal, and there is no ultimate survivor (unless, as Stephen Hawking and apparently Musk conclude, we colonize space-rocks virally). That is, there is no cinema, or there is only cinema. This makes the stereotype film echo something of utopist critical strategies – with apocalypse and utopian time both linked to still Christological memes. And the more it wills to repeat these redemptive futures, the more the acceleration of extinction logic surges – even when guided by the new version of Benjamin's stupid angel, or its hyper-industrialized front today, the Corporation. Headless, the Corporation is the head; incorporeal, the US Supreme Court insists it is a legal "person"; fictional, it nonetheless drives and, in effect, decides the real.

III Trauma-Type and the Cin-Anthropocene

> In terms of the animated image, we have yet to leave the prehistoric age.
> —Bernard Stiegler

These films have a reassuring effect. They allow the public (let's call it) to get used to these ideas while, at the same time, derealizing them. How, after all, do you depict something that does not happen at once but over scores of years and more – or, how do you represent species extinction, without a survivor to tell the story or regenerate another chapter (or imaginary future)? Moreover, how does cinema itself do that, separate itself from the human, take a picture of the Anthropocene (and mark its closure), and do so at a time when the death of cinema is in state (at the hands of digital transformations of memory)?

One problem with the term "anthropocene" is that, while it announces human mastery over earth and nature, it marks its disappearance. One can only confirm such categories after they pass, from another's reading perspective (including another species if there are no humans around). Cinema of the "trauma-type" puts it all into question by identifying outside of the human, against it as a figure in its frame: it turns against Hollywood man, what we can call his

cin-anthropocene episode – from the cave painting 36,000 years ago to the hyper-industrial die-off "he" confronts irreversibly.

There would be a torsion between photography itself, and cinema, and ecocide, since whatever the lens captures (or cites) is incorporated for use into the archive (is marked as dead, undead, or in its afterlife). In order for cinema to mark or explore what ecocide and extinction entail, it must perform a kind of cinemacide – particularly of the Hollywood model. It is a mode of attack, it is without (human) survivors. It does not try to evade, deny or forestall the "catastrophe" to come. It recognizes that "we" are the catastrophe – say, for all other life systems on earth, the sixth mass extinction event – and that we are in the middle of its unfolding. It breaks with the conventional attachments of the screen: identification with characters or faces, the projection of affect, the renewal or marriage that Hollywood drains human cognition with.

I will consider one counter-example in which the closure of human life is performed without remainder – and that as a suicide of cinema itself. Moreover, it won't be a disaster film at all, but starts as a social soap opera about a nervous bride getting married, who is a depressive, a melancholic – I am alluding of course to Lars von Trier's *Melancholia*, which also names a small planet hidden by the sun that, by chance, circles out to impact with Earth, obliterating all

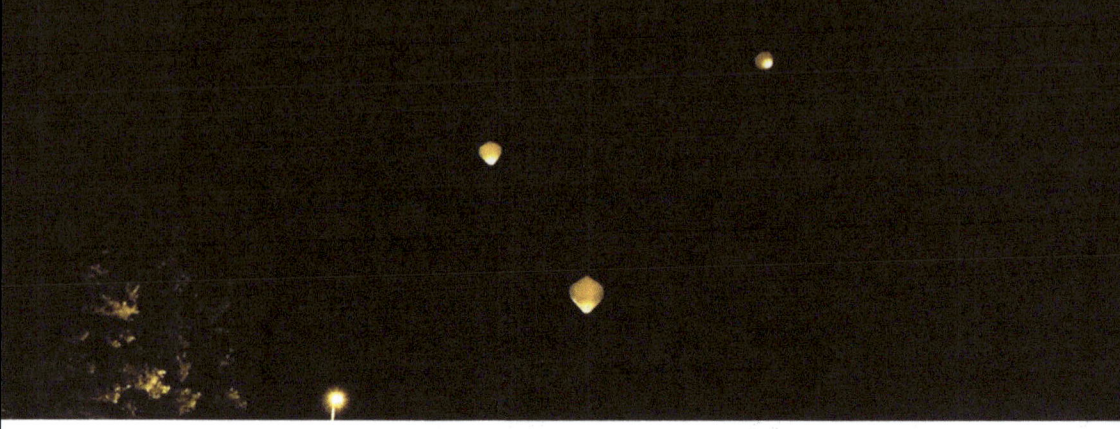

life. Kirsten Dunst's Justine withdraws from the entire social order, which she judges as evil, identifying rather with what is outside them – what is outside of man, and outside of cinema itself. The film is scored alone with Wagner's *Tristan*, which channels not only high Romanticism (what Baudelaire called the "acme of Western art") but implicates it, or a misreading of Romanticism, in the carnage itself. Moreover, it requires identifying against anthropo-narcism, the necessity of a narrative of survival, rejuvenation and self-witnessing being tacked on (since the "Anthropocene" posits a problem in time models we inhabit). The title names the planet and immeasurable material cosmos of spiral galaxies and rogue orbits, beyond this fragile orb. But it is the name, "Melancholia," coming from blackness and preceding "light," that names the film itself, its material, allo-human, trajectory as well as this particular performance. We see this, for instance, when lanterns become tropes for cinema – lighted balloons with writing on them sent up into the dark sky, by elite, dawdling humans at a marriage (that will be preempted): being reversed, at the limit of the lens, into gargantuan and gorgeous galaxies that obliterate anthropo-centered pretense – even as the little planet that comes from "behind the sun," which is to say before light itself is artefacted (the sun, too, is a technology), is a mini-double of earth, the closing of a circuit evident when the two orbs impact like the circles of a projector, or two wheels

or eyes, imploding: cinema turns against the human perspective, and itself. This is on display, for instance, when the wedding party sends up candled lanterns into the night sky – covered with scribbles and signatures, trope of cinema as a votive writing to the universe – only to have the screen answer with the overwhelming visuals of galactic star clouds, indifferent to the human ritual.

With this we need to remind ourselves what cinema is and has been, how it has been blended into what we see and not, into memory programming, in globalization, techno-genocides and weapon advancement (drones), nuclear fission, and so on. Bernard Stiegler posits, in this sense, an arche-cinema, older than Derrida's arche-writing. Going back to the cave-paintings, before writing, it established the organization of perception and consciousness in the projection, by light, of marks and inscriptions, onto the wall of a perceptual commons. The era of modern cinema was a technological flowering of a model of memory and cognition (indeed, Stiegler will say, of "life" itself) that we still inhabit, 36,000 years later – and which the death of cinema today before digital transformations brings to a close. The model echoes in Plato's allegory of the cave (again, a cinematic model), as it does in the "consciousness industry" of Hollywood and telemarketing. From this perspective, we might speak of a cin-anthropocene era, which, however, would also bind cinematic processes to the archiving, capture

and consumption of Earth's life forms. To have entered the photograph is to be, virtually, dead, targeted, about which much has been written, and which is why Benjamin made the photographic image a cipher of our suspect practices of mourning. In a distinct way, Justine breaks with mourning, and affirms ecocide. Thus, the stunning scene where Kirsten Dunst is viewed waiting for Melancholia, this gorgeous giant sphere that annihilates, like a lover – no man will do – in the pose of a pre-Raphaelite citation. The film builds an array of citations dragging in the totems of high Western culture, like Wagner himself, which inscribes this annihilation in the practices of identification, mimesis and mourning which cinema seduces with.

And it also cites the one film in the canon that engages fully with a revolt of arche-cinema – Hitchcock's *The Birds*, in which, you recall, swarms of flying black wings attack visibility itself (and the eyes). This occurs as the two sisters pick and collect blackberries silently: there are many little black pellets gathered, like so many tiny black suns (inverting the giant galaxy shots), but these may become the one of a bird crossing the white sky – before ash-like flakes fall from the sky instead.

This cites a famous tracking shot from Hitchcock as his blond character, Melanie (again, the black name that becomes melancholia), waits outside the schoolhouse as birds gather on the jungle gym to attack. We can call each "bird" a hypermnematic slash.

The Birds is channeled since it too, a half century ago, a work of the Anthropocene – in which the birds, identifying against humans and human perception (they attack for no reason, accelerate, and win, driving the grade-B actors out of the house, the screen, the narrative). These birds are at once marks out of which memory and cinematic representations would coalesce – but here they attack a diseased order, a spellbound town. It is not accidental that these birds are linked to oil, what we call "stored sunlight," without which there would be no artificial light or cinema, yet itself black (and co-responsible for hyper-industrial auto-extinction). It derives from these birds' prehistorial forebears (dinosaurs) and fuels transport, cars, speed, the screen itself. And, you will see, it is called "Capitol Oil," drawing in not only "capital" but the head (eyes, sight, cognition). "Animation" itself belongs to arche-cinema – genetic engineering assumes this – and Hitchcock's birds testify, like *Melancholia*, to a blackness or technic that precedes "light," "Enlightenment" tropes, the flickering screen.

Thus, *Melancholia* works by suiciding cinema, and marking that as the artifice it is – no disaster film, it takes place solely on an allegorical, gardened super-estate at once of Euro-nobility and American advertisements (and TV's Kiefer Sutherland or *Spiderman*'s Kirsten Dunst). Unlike post-apocalyptic fare – which has a certain link to utopic criticism, in requiring a redemptive narrative – *Melancholia* shocks the illusion of "shock," but also inscribes cinematics as the agent of the annihilation, looking at human life, as Justine does, as a criminal disaster in the form she finds, and withdraw from, it. But that is also what we are implicated fully in by Wagner's score, by the affect and seduction it accomplishes – as in the crying panics of Claire, Justine's sister, at the "end."

When Justine marvels at electromagnetic streams rising off from her fingers as from pole wires, the screen films and pleasures in the drawing off of all from the screen itself, energy and "life." This emergence in high Romanticism is kitsch, as well, marking us as stupidly inscribed projecting pathos and mourning (which is always self-mourning) on the screen. There is no escape – no arguments – and Justine assuages the futureless boy that his (now suicided) father, Kiefer Sutherland, forgot "the magic cave." This bare figure of the cinematic structure humans conjure and dwell in, film itself, is a bunch of branches like a teepee (Native American cipher), but also like the bare structure of the jungle gym Hitchcock's birds assemble onthat is,

lines, converging parallels, a "magic" nothing but the structure itself, and the mega-disaster, nothing more than billiard balls for the night sky, erases the screen, closes out the cin-anthropocene – which Justine authoritatively says would not be mourned or missed.

Melancholia is without exit – irreversible, as the planet given the film's name, what Benjamin called a "one-way street." This is dramatized when Claire runs around the grounds in a panic carrying her son, as if there were somewhere to go. Its ruthless logic coincides with the emergence today of a critical positioning, linked to the "Anthropocene" (or, recognizing China as the key civilization in Earth's history and determinant player today, the Sin(o)-Anthropocene?). This position abandons utopist thought as circumscribed by a moment that is now closing, which is inadvertently linked to the accelerated extinction itself: first, by viewing history as social actors, a matter of otherness, power and progressive justice, they have occluded the vast allo-human primacy that now breaks this spell; second, as the last echo of Christian salvation, it bears within it the apocalyptic DNA that goes into the Hollywood blockbusters, projecting a survival and renewed narrative.[54] To think, now, for after the catastrophe (which can be dated forward or back, since we are in the middle of it already), is simply a given, rather than not seeing, denying, trying to put off, or waiting until it seems past – part of the filmic archive. North America continues to experience a phenomenon called the "polar *vortex*," in which climate tectonics reversed (cold and hot), with NYC dropping in temperature 50 degrees in hours. The affirmation of ecocide and its critique of "utopist thought" is unthinkable, though, without cinema's turning against we "humans" (I exempt myself from this category), which it had always quietly been – since, platforming cognition, it has never been other than sheer technics. If the era of climate chaos would be that of cinema, the cin-anthropocene, it is because the lens has always been predatory of what it shoots. If zombies are now normative, is it because consciousness or "life" is experienced as a circuit of mysteriously pre-inscribed memes one has forgotten are not, themselves, organic or even "alive"?

IV Disorganizations of the Inorganic

> ...the cinematic pharmakon as art is what makes it possible to struggle against the cinema as toxic pharmakon, i.e., against that which enables the play of the traumatypical secondary retentions and protentions of psychic individuals to be short-circuited by reinforcing their stereotypical secondary retentions and protentions.
>
> —Bernard Stiegler

So, a small thesis: in displaying certain Hollywood variations of these logics – from *2012* or *The Day After* through the zombie apocalypses and post-apocalyptic tourist films, in which we somehow are still there to prevail and witness our, usually, cannibalization and re-emergence (even at the end of *The Road*) – one does not find an expression of cultural anxieties, a "climate-change 'unconscious'" diverted and marketed into mass fantasy. And one does not simply find, sometimes, as in *Melancholia*, cinematic logics turned against the human, the viewer-consumer who wants to identify with the face on the screen still. Moreover, the disaster movie is not only used to get the public imaginary familiar with these logics. They now seem normal when they arrive, if not planting a desire to see and partake (they've seen it in the movies already). "Shock" is normative, and hence not. One might suggest that the Hollywood blockbuster or its affiliates (*Avatar*, or again even *The Road*) ensures that these logics are not dwelt on, since they will have been stored as managed possibilities. It is, after all, like magical warding off – and one can turn to another channel somewhere. Only one cannot.

But there is the exception – what Stiegler calls cinema as "traumatype."[55] *Melancholia* pulls the ripcord on this cinematic contract altogether, performing at once a sort of death of cinema and a closure of life without remainder – a sort of suiciding of cinema, a cinemacide from its afterlife, or a suicide of and by what might be named the cinanthropocene (a feat only cinema can do). It is an act which, like *The Birds*, breaks with the viewer's projection of affect and even artistic wonder on the screen. Another terrestrial body will impact, no one will miss us – or read us afterwards – and there is nothing apocalyptic or post-apocalyptic about this. It is as random as the dinosaur strike but woven as if a judgment. Unlike the other films or their trends, there is no survivor, no re-beginning, no archive: it is the closure, with

the screen, of the Anthropocene tout court, which is the latter's premise (as a category that can only be applied after the disappearance). But that has been implied since there was arche-cinema, since there was photography. The birds win: they drive the human stars off the screen, out of the house (eco), into a sort of blank exile – but what is asserted is not nature winning but the technical premises of visibility itself disowning us.

Now, I leave it to you what this has to do with that other movie that we mentioned – the hypothetical and very bad Hollywood script 20–: The Un-Movie." That is, what is given us as a bad script, but which may in fact be the unvarnished basics of a situation that is only catastrophic from the perspective of "life as we know it," or knew it (once). The timing of the catastrophe has always been up for manipulation: it is coming, we can avoid it; it is coming, we can sidestep it or adapt; it was already here, it has no one time, it was there from the start, and we survived it (even as zombies), and so on. But *Melancholia* has a different point of view: we are fully in the middle of the catastrophe, and it has been long determined, as we glide past tipping points (as numerous official reports now advise us, boringly and to no response). Moreover, "we" are the catastrophe – if viewed from the perspective of any terrestrial life form, or, for that matter, the galactic neighborhood. This is why *Melancholia*'s paper lanterns, reminders of the fragility of cinema, dissolve as the tele-scopic view is reversed and the gigantism of billions of stars obliterates "us." And it is also why, at the close, they huddle in the "magic cave," which is the tracings and inscriptions of arche-cinematic culture, the same representational conventions that led us into this exitless place, this *atopos*, which is no longer an "Earth," and which is in mutation (like us), which we find ourselves in today: "the Un-Movie."

Personally, I would turn down the script as clichéd. But then, that is where we are: disowning the real because it feels (and is) Hollywoodish.

Who, after all, would believe a "super elite" (it is not my phrase), globally, have embarked on a species split – or that you are the new "proletariat" (or "pre-cariat," as they now say), funneled into service jobs or harvested for megadata and, occasionally, organs, a disposable population? Who would believe that, all along, during waves of corporate-sponsored denialism and faux science, "they knew" – as many not so deep state reports, including those of Big Oil, confirm in fact (NASA, IPCC, UK defense and CIA reports, and so on). Or, rather, who would believe in the finality of consequences? Moreover, who among the intelligentsia, variously defined, would acknowledge the choice: either give up the future generations' options, and eventually their existence, or give up your own twentieth-century conceptual investments and representational "truths" which got us here (you can even keep the SUV)? Certainly, as regards response, a certain tradition of the "utopic" now stands in that regard as much as what was once called "capitalism": that may be the shock value of Hedges's impasse and revelation. A certain battle, in any case, seems to have been decisively lost, already (that is what "tipping points" mean, if passed), and to the degree utopist politics finds itself suspended – made possible by a certain phase of stability and growth in the twentieth century, like democracy – the question is where, and how, the anthropocene requires us to give things up.[56]

If even the best of Hollywood negotiates with the stereo-type – and is usually designed as a product with that in mind – the former parallels the utopist strain of critical ornamentation.[57] The *trauma-type*, by contrast, destroys these to be shed as packaged streamings, determined still by the anthropomorphic program (here, that there are first and only power, justice, competing groups, social structures, economic pillage or "capitalism"), and seems to leave nothing in their place. That is untrue. Today, the *stereo-type*, like the twentieth-century imaginaries we inherit, feeds the acceleration, is entirely correlated to the accelerations of the cin-anthropocene. What seems harder to comprehend is that that itself is the zombie. As Stiegler notes, cinema is now "dead." In terms of Benjamin's allegory of the angel of history destroyed by the storm – the question today is moving from the back-looking angel's position, wanting to make whole (utopism), and thinking from, with, and within a storm. Moreover, the problem seems to be, with us, that if there now is no escape, there is also no way back to earth: we are not in a utopia certainly, but a dystopia is not right either – in a way, "we" have never been better off, for a certain "we," and for the moment.

Rather, first of all, atopia: Sandra Bullock, floating in deadly space junk, without up or down, without Earth or non-Earth, without gravity (the Oscar-winning *Gravity* (2013)).

One might call this perspective less "cynical" than "cinecist."

If the activism of "social justice" is not altered at all by this – which does not matter if civilization has years or centuries left, so this fear of cynicism is a canard – the trope of "social" justice, between humans, the supposed fulcrum of "ethics," is contextualized by the parentheses of time that supports it, like the post-world war US economy or the Little Ice Age's retreat. The sliding orders of "justice," if there are such (hint, it is not "deconstruction"), risk turning against a certain species – if it is one – in the name of "life," "Gaia" (whose revenge is promised by Lovelock), or of extinct species. This is what Justine experiences that is named "melancholia," a sort of view not through the trope of "light" or "the Enlightenment," or the void corporate ruins of the "1%'s" estate, the last-man culture she defaces in seeing (or is first defaced by). I will return to this event of defacement, since it is precisely not a figure, by returning to the problem posed at the opening – that of a "proletarianization of the senses" that, rather than be a struggle or a plague to war against, has proven too successful: in particular, the sense which is not one, the artifice of sight, the eye, or reading, all of which would have been initialized by arche-cinematic technics.

V Polar Vortex

> Our time is a time in which the mafia and the oligarchies remorselessly chase out the bourgeoisie – a bourgeoisie that, although it is philistine, is still too cultivated in their eyes.
> —Bernard Stiegler

The recent *present* arrives not only with a suspicion that it's "too late" – but with a storm, specifically the polar vortices that reverse poles of heat and sense. It also just happened to disclose the remarkable coup of a hyper-elite's massive wealth transference, while climate "denialism" sufficed to paralyze and forestall focus long enough, apparently.

What we have seen is that a certain "proletarianization" and disempowermentwould have been engineered so completely, and irreversibly, that one had to distinguish it totally from past versions, such as Marx's. This time it would not be about suppressing labor or labor costs, not while entering a period of rapidly advancing robotics in which unemployment would be normative. This time any "proletarianization" involves something other than a class – but a disposable population. [58]

In terms of Stiegler's pharmapathic war on dis-individuation and adaptation or subsistence, there is and there is not a road bump. If a battleground is lost, one pulls back and repositions. Stiegler uses Simondon to engage the socio-mnemonic and libidinal economies in default (the "disaffected"). His Simondon is one that is "supplemented" with Derrida and Derridean technics (who is effaced into Simondonian process). In positing new "we's" for this moment, to counter the short-circuiting of time and *attention*, however, there is a superficial mimicry of the utopist, restorative, hopeful gesture that seems wired to the genre of political endgames.

Stiegler's work and position is indispensable today, as he traces and integrates the necessary backloop through perpetual technogenesis of an encompassing "transformation" – which requires a retracement through arche-cinema, prehistories through which archival programs are initialized, the default of epochal ills set (or erased). (In this, it recalls in minimalist terms that Derrida's invocation of the Platonic *khora* – and it is Derrida's reading of Plato's pharmacy that Stiegler would precede.) We can say there are two ghost columns within this side of Stiegler, Simondon and, often effaced, the Derrida that supplements the former (impossibly, but that is its positive condition). In this domain of Stiegler's production, there is the illusory narrative of a history of progressive exteriorization with the advance of the technologies (and, specifically, mnemotechnics) that produces us – but also, in our hyper-industrial acceleration, ecocide today. The ill or *grand mal d'archive* is manifest but invisible to itself, and Stiegler's writing at all points operates as a counter-toxin. But a year of polar vortices and of covertly conceded auto-extinction interferes with the Simondonian promise of a "transindividuating" we, consolidating, adopted, fused in the reinvention of care and "new technologies of the spirit."

Let me take advantage of this *hiatus*, these polar reversals and blinds, to play contrarian – and separate out, in the name today of a black enlightenment that I would dare to call, in part, Stieglerian without "Stiegler," and which is quite different than that Stiegler cites in Husserl or Valery, when taking in Europe's self-obliterating twentieth-century wars, total wars. (As noted, it is interesting that the motif of eugenics, supposedly put to rest by the Western market democracies, should re-emerge so blatantly out of the latter, now dependent on wealth alone and not, say, "race.")

I will attempt a *cynecistic* reading – one in which the pre-individuated premise of arche-cinema is in attendance. A cynical position, only in the abandoned sense in which that is misapplied to the ecocidal thought today ("not cynical" means not to give up "hope," which is exactly what, say, Latour recommends as a premise for moving forward – together with the word "future" – which may be, superficially, why Stiegler accuses the later of cynicism). Any term of banning needs periodic inspection, and it turns out if we return to look at, say, Diogenes, we find a very different force. Diogenes – who knew Socrates, harassed Plato (and dismissed his "Socrates," while being called by Plato a "Socrates gone mad"), who dismissed Aristotle and told Alexander to stop blocking the sun for him – this Diogenes, among these generations, went about with his lantern in the daylight, looking for a man, an "honest" man (in any case, not a "featherless biped," in Plato's unhappy evasion, unhappy after Diogenes showed up with a plucked chicken, ecce homo). Do you see the point? Diogenes witnessed the entire conceptual edifice and programmatic of logics assembled and installed – then its transposition into the conquest of a world in Alexander, and this was his response: absolute suspicion, attendance at an event run by nerds and cons, caught in the mimic curse of knowing Socrates (but without writing). Before there was "metaphysics" officially, before there was the sun as father, good or virtue quite (formalized in Plato's engineering of *eidein* to fuse knowing and seeing, cognition and the eye, light and knowledge), Diogenes spits the dummy, so to speak. Why? The progenitor of the pragmatic and cosmopolitan stoics, with their dissociating techno-consciousness, may have been a performance artist, but wasn't, well, a cynic exactly (canine references aside). He practiced defacement before "face" had been installed fully, but he also doubted currencies, the communal investment in mutual coinage – hence, his act of corrupting the currency which earned him initial exile. But it is that lamp that bugs me. Let's say, for a moment, that he bore, in this self-canceling double positive of "light," which does not disclose the sought in any case (a "man"), the cinematic torchlight out of the cave or enclosure, which Plato would in fact consolidate and codify in his allegory of the case (which, Stiegler reminds us, echoes, disqualifies and forecasts the cinematic apparatus of "consciousness"). The lamp is lit with oil, a techne-light, the assertion of a technic at the core and before "light" (including solar), which shows light to be not light (not the source of light, or vision). What did Diogenes know, as he turned away from one way that the Western DNA was about to be artificed and set, even if inversely – as if he knew, from proximity to its architects, where this was going, even before the boy wonder showed up as an umbrella, grinning. Would the *eidein* itself be, for him, a sort of propaganda trope, put in cynically by Plato, to shape the polis (and run the projector) – the same Plato who, it seems rarely noted, deconstructs the

Idea in advance of its circulation (has Parmenides do so, in fact), what would be inversely imprinted as his signature or brand product by the hermeneutic community? Plato still conducted, in the mode of *dialogos* without the authority of intentionality or codified logic (Aristotle's grammaticizations), a pre-individuated or arche-cinematic writing scene and, yes, pharmacy. (It is not incidental that Stiegler, uniquely among contemporaries, discreetly chooses to channel Plato, triangulated again with Derrida.)

Diogenes's lamp in the sun echoes *Melancholia*'s cinemacide at the far end of the trajectory. But this premise too lies at the core of Stiegler's technogenesis without reserve: "light" itself is the effect of technics, intervals, waves, fission, fossil fuel as black "stored sunlight" today, life cannibalizing its own anorganic waste. "Light" is black, much as the trope of coming to light, as if to sight and to cognition, institutes an awkward metaphysics of whiteness. The chasm of "inequality" and the rise of a financially engineered and proactive survivor caste separating off and setting up passive prolonged die-offs, are not the usual products of "corruption" in the ethically tainted and criminal sense but of centripetal anomie once tipping points are breached (hence the Trump-Musk acceleration in leveraging this irreversible acknowledgement). Nor is it accounted for as a perpetual end-running or "short-circuiting" that rewards the com-predator directly – the outbidding of and by perpetual or pre-originary theft, which informs *technics*. It confabulates outside the law, today, as it seems the pure reflection of the norm itself, that is, of exponential technological advance as end-running and short-circuiting, as taking the previous mode of "tertiary memory" (or inscription) and gaming it. The rise and arrival of A.G.I. feeds off the pro-active vortex, not incidentally sucking in gargantuan energy demands that just happen to coincide with the Trumpocene's "return" to oil, as cascade events trigger. This is what gives the ".00001%" their stupid innocence, subordinate to the incorporeal Corporation as new (headless) head – the command of "exponential growth." If the perpetual death of cinema closes the circuit back at the installation of arche-cinematic templates, out of which sight and movement would coalesce or be set (together with animemes, the model of the hunt and mimesis), and so closes out the cin-anthropocene, turns on it itself, it returns the sociopathic CEO and kleptocrat as cultural hero, along with the serial killer and cannibal, to the initiating act of technics itself: pre-originary or ante-Promethean theft, pyrotechnics. That Diogenes bears the mute image of pyrotechnics in the daylight and cannot see a face coalesce before him also forecasts, virtually, a future dependence on "oil" for that light, and with it, the cycle of ecocide it discloses. Diogenes represents an alternate assemblage of histories by forestalling the imposition of face, mimesis, Aristotelian logic, empire, the "Enlightenment," and the hyper-kleptocracies that could not not accelerate into, essentially, an innocent mime of pure technics itself. The *Autos*, fabricated on stage and in dialogic genres (without authoritative "I"), used to anchor the legal subject, the psyche as one, interiority as eating and property, arrives in the hyper-industrial or ecocidal era as exposed, a technical algorithm, a bare or minimal effect of an accelerating, addictive cycle (one about to flair in the last, frenzied era of resource extraction, e.g., the polar regions).

The reason this interlude is relevant, this backloop, is that it opens a different reading of why Stiegler's strategy, on the Simondonian side, may be bifurcated, or why the "proletarianization of the senses," a lost battle, may recede as the *hypomnemata* themselves emerge as the scene of contest. In Stiegler, if the "senses" and perception are efficiently captured, interrupted and impoverished, having "savoir faire" outsourced and not returned, leading to the society of "disaffected individuals," it may be too automatic to assume, with Simondon, that the "transindividuated" "we" can be crafted or artefacted, adopted and given "new technologies of the spirit" to prosecute what is essentially a war over ecocidal totalization – which, again, the emergent politics of extinction promises, clearly, by normalizing "adaptation" and "geo-engineering" and mnemonic subservience (implants, surveillance). Stiegler tracks the technicity of grammatical positions – the artifice of the "I," the "vile desert" of the impersonal and feudal "they" (as in, "they know"), but while noting the "we's" non-existence, he cannot but extend its import as and in the experience of "con-sistence." Voice cannot reposition itself without such a *koinos* (an-archival setting). And yet, to fully anachronize this today, as any glance at Fox News mantras demonstrate, it is precisely the technic of the "we" that has been most easily simulated, botted, telemarketed, affectively shaped (as on the screen) and capturable in advance of its (non-existent) arrival. Stiegler chooses not to close this window or trope – it is necessary for him to finish the grammar of the sentences he begins which index a transformation they themselves testify to performatively. Yet since this "we" has difficulty escaping the illusion of a human-on-human collective, which in turn authorizes and re-initiates, one could say that the only possibility of a communal force and experience, today, lies in retiring the "we" itself as a grammatical trap – stepping outside the anthropic screen of Hollywood, face, affect. It may be that, in the midst of his Dantesque survey of the techno-psychic underbelly of hyper-industrial society, and directing a prosthetic restitution of care otherwise, or attention, one should first accelerate disaffection, insofar as the "affects," like stereotypes, are the most citational of replays – rather than an ontological immersion of world-psychic being (or its wreckage).

VI Hyper-Matter and the (Once) Coming Wars of Re-Inscription

> Our time is a time in which the mafia and the oligarchies remorselessly chase out the bourgeoisie – a bourgeoisie that, although it is philistine, is still too cultivated in their eyes.
> —Bernard Stiegler

Arche-cinema as the technic of technics precedes the metaphorics and artifice of "light" (or Enlightenment protocols). If, as Chakrabarty painfully notes, the figure of "freedom" and a universal Humanity, like dialectical resolution, advance ecocide in today's back-glance – much as market democracy guarantees it (Arundhati Roy) – then the Enlightenment as such would be a sort of black ops, much as it rhetorically covered the predations of the colonial period, or found its protocols both muttered and then defaced by Mr. Kurz. But I return to the "we" – which, in structure, is little different than the "I." And, despite "voice" having been pitched iconically by Plato as porporting a presence before technics (writing), the "voice" would already be a techne of (script) dia-logos, a mnemonic citation. It is "one" which the "I" not only can disown (as in Euripides's Orestes's claim that his tongue spoke, not he, as if it were read from a script by another) but which is constituted in the re-citing. It, *voice*, "I" was always the product of a mnemotechnic recording device, one traversed by others' voices and mimicked words. The *Autos*, the self and the same, was automated, and when today – at the close of the cin-anthropocene – Wall St. accelerates by "short-circuiting" and rigging or end-running, and corporate personhood usurps the place of the head, and when hyperbolic and outbidding corruption and mafia circuits are normative, and the planet tips irreversibly into ecocide for this phase of terrestrial life, it can seem what is expected from a being, "species," organism, hominid, as you like, that is being hollowed out or exposed as the invention of and by technics. From this cynecist perspective, if this were only a Western narrative and if the "Anthropocene" did not index itself to the Aristotelian *anthropos*, one might speak of the circuit that closes out today, like that of cinema, ending as a sort of grand theft *Autos* (by itself). The cinematic lens, or "eye," was always predatory, the eye that hunts, targets, seeks to eat, mummifies or "kills" in shooting or archiving.

It was, it turns out, irrational for scientists and ecological Cassandras to imagine that, were evidence presented, a self-preserving leadership or "humanity" would address the crisis or try – the way Hollywood assumes some collective mission might rise were an asteroid getting too friendly (Hollywood, Bruce Willis, check), or aliens, or event zombies. Such a collective ("Humanity") would never arise aside from corporate campaigns of disinformation, political suppression and the inability of it to generate aura or even develop a serious vocabulary. On the contrary, having put up the memorial marker of the Anthopocene, having posted it, having taken pleasure in it as a trophy, disappearance is implied (if not perversely desired). The attempts to shame with guilt, in the media, as when CNN ran a piece on oil titled, "We Were Warned," are greeted with mockery, irritation or guilty pleasure – once making "Al Gore" the unfortunate poster boy of contempt ("inconvenient truth"?). *Man would rather will nothing than not will,* someone muttered. The word "anthropocene" is like a photograph, condensing a geologic scarring to a signature. But, like a photograph, it is aimed not at other human readers but at an eye, a recognition from without, which is to say after its disappearance – a reader that is not guaranteed, nor guaranteed to find it readable. It functions like art, the agency of the "hypermaterial" that is not tied to any organic life, and is not unrelated to a "consciousness" that thinks itself living but is the product of virtually cinematic process. Cinema, says Stiegler, is "life." *Cin-animation*. Not only "consciousness" but "life," which does not exclude cell production, genetic coding, body formattings, abio-semiosis.

And this backloop is where "Stiegler" departs from, where "new technologies of the spirit" must proactively be launched, within sentences that traverse epochs, open or mark others, posit a backdated "transformation," assume responsibility for the full circuit that is now exposed, for initiating new inscriptions.

If "dis-individuation" has advanced and the "proletarianization of the senses" been efficiently taken, Stiegler has the nuclear option in what he correctly sees as a war: he can turn to the *hypomnemata* themselves. It is here, again, the arche-cinematic returns. *Hypomnemata*, which can be translated as "inscriptions," are discreetly handled by Stiegler, since they have no phenomenal access, but they are identified again and again as the prerequisite of any mnemotechic regime, archive, as if the mechanics of "transindividuation" can now be hacked, or simulated; if the "senses" have been put in a spell, impoverished, or mnemonically captured – so that, say, the eye cannot see what is right before it ("climate change"), or if the "we" sought is the NSA's backdoor portal in effect, Stiegler has the option of going nuclear. He does not, like the American Hedges, gesture to stepping outside a totalized system that accelerates extinction.

The *hypnomenata* do not entirely correlate to the celluloid of the cinematic template. As material orders of technics – requiring marks, sound, cave walls – they reside at the disappearing point of the "organization of the inorganic." Since aesthesis, with its product, "perception," is generated from and within this non-site, and since tertiary memory as supplement pre-inhabits (and divests) any phenomenology, these *hypomnemata* format the hardware, select and preorganize the visible, produce tropes and descriptives explicitly designed to efface or deform them. Since they have no mimetic form, are not parts of "meaning" and retain agency that is incalculable and subject to mutation, to invoke "inscription" or *hypomnemata* does not deliver some object of attention (attention departs from these, or not). This is why all of Stiegler passes through the performative destructions and perpetual reconfiguration (and renaming) of what should not quite be called a "politics of memory without a *polis*."

Arche-cinema is not a political "we" or organization. If the corporate state has not only penetrated a collective mnemonic order (which is nothing new), but muted the always prosthetic "senses," what they would aim for next is the inscriptions themselves. Parodies of this play out daily in telemarketing streams and geopolitical wars. Since these issues now rub against extinction accelerations, or have passed them, there is nothing lost by pointing out that Paul de Man called this order of hallucinatory perception "aesthetic ideology." De Man uniquely circled this thankless zone, called that of "material inscriptions," with just this contingency implied – how do you name or analyze or displace a totalization (this is what Hedges accuses the corporate state of having succeeded at, and what Stiegler performatively wars with)? *Aesthetic ideology* is like the screen of the stereotype of Hollywood film, and is the domain of mimesis, identification, face – contracts which popular cinema or TV launder and reinitialize daily. It is, writes de Man in different permutations, the confusion of phenomenality for the effect, or evasion, of inscriptions (which cannot be phenomenalized).

When the titanic, Black and illiterate "Rider" (reader/writer) crosses "the junctureless backloop of time's trepan," Faulkner conjures the performative hyperbolism of reinscribing temporal circuitry. This gesture haunts modernism and informs Benjamin's non-existent "materialistic historiography." It is routinely evoked in Stiegler's sentences as a "transformation" that is projected as already in place. I have suggested that if Stiegler is a writer of the post-Anthropocene, he is also writing for after the disaster, which can be fore- or backdated indifferently – since, strictly speaking, within the default that spawned these techno-histories, "we" are (is?) the disaster. The problem of "social justice" gets complicated here, if the "social" as named or produced, circled off, accelerated to spawn accelerated mafia, "short-circuiting" and taking the preceding state of tertiary memory (or technologies) as the target of acquisition, outrunning, citation. This makes for the indecent impasse Latour eyeballed when his attention turned not to the scandalous subtext of the "climate crisis" (self-extinction, complete carnage of current "life"), but communications, the game of denialism and parsing of "science," the enmity of or within language. One awaits a necessary dismantling of the very vocabulary in circulation (ecology, green, sustainability, global warming) – rather than just, as Yale's center of climate-change communication matters informs us from taking a poll, that "global warming" has more traction than "climate change" because of the implied heat, a telemarketing or branding issue. What is becoming apparent, however, is the unexpected locus of the disaster, to which carbon displacement, biomass, hyper-financialization, and so on, are linked, and what is at the heart of Stiegler's transposition of Derrida's "politics of memory" phrase. What he means, though, is mnemotechnics, which leads to *hypomnemata*, which evokes the khoratic logics of arche-cinema. The drifting fortunes of "deconstruction," its use, orientations, ethical collapse and priestly rites, are not unrelated: removed from circulation, thanks in part to Derrida's auto-archiving preoccupations (worry about legacy, or anything at all, surviving him), absent largely from the so-called climate debate, if there were one.[59] Some have recently written that, for him, de Man was also placed in that zone – each of which withdraws from and defaces the archival premise that there was a "metaphysics" to deconstruct. One might quietly tag Derrida for this omission today as costly.

To say that ecocide may find its "cause" not in "the great acceleration," but back in linguistic memes, codings, void citations and the shared reflexes of hermeneutic switches, mnemotechnic swarms and the crisis in reference which climate chaos opens, is to say the obvious, as when some point out that today's consumer-citizen may have no model, no experience, no memory of the disruption of some "norm" (the status quo artefacted as natural state), since its "senses" cannot cognize what it has no inscription to recognize – a technical alibi dismissed by the arche-cinematic.

On De-Extinction, AI, Cinema without Screen

Hypothesis: That cinematization had all along been the forerunner of the rise of A.I., and that Spielberg's A.I.: Artificial Intelligence, is compelled to ruminate on cinema as de-extinction and as what Derrida called (speaking of cinema), the "absolute simulacrum of absolute survival."

**9 Did an "Anthropocene" Even Take Place...?:
Notes on an Image in Spielberg's
*A.I. Artificial Intelligence***

What that belt of swirling storms girdling the planet's midsection says is that the climate is not changing; it's that the climate has changed. None of the countries on the map are undiscovered. The storms show exactly where we're headed, and which parts of the world may not belong to humanity any longer.
—Adam Rogers[60]

The film [*A.I. – Artificial Intelligence*] is about the robotic post-human, and it uses a technique [CGI] that's occasionally described as "post-cinematic." Am I weeping for the death of David's mother, for the death of humans, or for the death of movies?
—James Naremore[61]

Cinema is the absolute simulacrum of absolute survival.
—Jacques Derrida[62]

Before there were films, there was cinema; the flickering shadow play of fire and motion on limestone cave walls.
—Darran Anderson[63]

I

Cinematics, appearing as sheer technics before any writing and as a template of "consciousness" production, might well be termed the communal locus of free-standing "artificial intelligence" – running from the conjuring of animation and kinetic mimeticism on cave walls, torchlight waved over lines and marks, to cineplexes, digital totalization and decoupling.[64] The resultant hive mind, now dwelling in portable screens, becomes inextricable from what is today mislabeled the "Anthropocene" – which is to say, inextricable from the arcs, today, of extinction. "Today," when we witness, as if we were watching a movie, dissociated, tipping points pass, cascading feedback

loops trigger. This figure of passing "tipping points" is peculiar, since it could be said to disable any "arrow of time," which boomerangs and contracts, and instead of opening futures, focus on delaying the vortex. Once said tipping points are in the back mirror, as they are in the opening image examined, according to the voice, there emerges de facto a discreet politics of managed extinction, which, it can be argued, subtends the present. The inside story of AI is inseparable from this mnemo-politics – digital totalitarianisms, "post-human" phantasms, species splits, geo-trash escape logics (ex-terran colonies).

II

Two imaginaries dominate this "today" – 1. that of climate panic (and extinction logics); 2. that of AI (and escape-extinction logics). The two rarely overlap in our discourse, as if two sides of a Möbius strip which do not touch and appear immunized to the other as they accelerate beyond reversibility. There is what I would call an "image of image" that seems to fuse these two and binds both climate chaos and autonomous AI to the arc of cinematics. It is a surprising image, since it seems to look back at the so-called Anthropocene, or *ourselves*, and speak from or as the screen, from or as if cinematics itself. What would a sentient screen say to us, after all, we captives of the "Anthropocene," if it dissociated itself from us, or if the POV had transferred to the sentient machine or robot entirely? The "voice" of the opening shot turns out to be itself computer generated by figures of pure cinema without us – a Super Mecha called "the Specialist." These Super Mecha arrive long after human extinction to retrieve the boybot David from the freeze of a sunken Coney Island. But more on that later.

III

The image in question is what may be called the elastic "shot" opening *A.I.* It is sheerly kinetic, a shot of the raging sea alone, stretched out for a moment to accommodate a voice seeming to provide a narrative setting, a backstory, yet withholding key information from us. He is reassuring, "godlike," straight out of British TV (Ben Kingsley), telemarketed for white suburban middle-brows, so we assume whatever backstory is evoked ("Those were the years...") ends well enough to have him as its reader. The voice mimes the missing logic of the term "Anthropocene," that it can only be spoken from long after our extinction, to become geologically readable. The entire relational artifice of voice to screen, and the who of the former, is in question. It references the era of "climate chaos" in retrospect, a given, keeping from us that it speaks from long after our extinction. The oceanic churn, generator of "life," precedes the human episode and supersedes it, spoken after we are gone and with no people in the frame. It precedes us and is there after us. The opening words of the film are spoken over a screen shot of the sea – source of organic "life" – by the calmly magisterial voice-over, of and for the screen itself, already titled, of "Ben Kingsley":

> Those were the years after the ice caps had melted because of the greenhouse gases, and the oceans had risen to drown so many cities along all the shorelines of the world. Amsterdam, Venice, New York, forever lost. Millions of people were displaced, climate became chaotic. Hundreds of millions of people starved in poorer countries. Elsewhere a high degree of prosperity survived when most governments in the developed world introduced legal sanctions to strictly

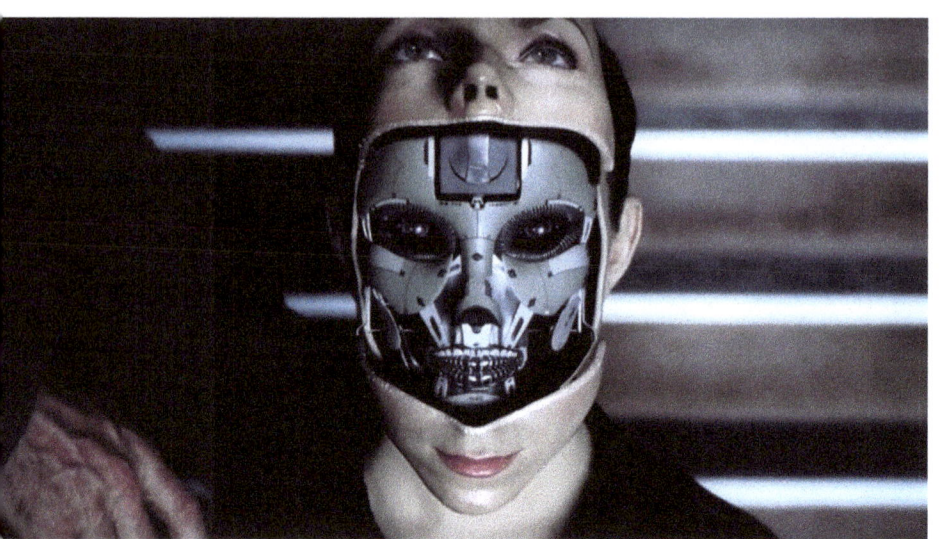

license pregnancies. Which is why robots, who were never hungry and did not consume resources beyond that of their first manufacture, were so essential an economic link in the chain mail of society.

The last trope, "the chain mail of society," is cinematic: the linked chain (recall the closing of *Psycho*) evoking early celluloid bands, the mail oscillating between an informatrix network and the weight of the defensive role it serves, entrapping.

IV

The film tracks the emergence of a first autonomous AI, David, a boy-bot adopted to replace a lost, comatose son. In the perpetual uncanniness of David's face, adoption, "imprinting" and then abandonment like an unwanted pet – he is released into the fairy-tale quest of becoming "a real boy," "unique," to be loved and returned to his "mommy" (Monica). The screen incessantly stages allegories of cinema, not through citations (primarily of Hitchcock) but in the escaped slave robots – marking, from the first introduction of the secretary Mecha, Sheila, that machinery (and CGI) reside behind every screen specter or "face." As David is apprised that he is but a copy of a copy, he plunges desolately into New York Harbor for two thousand years, during which the human extinction is now accomplished, warming seas reverted to a nuclear freeze for reasons not marked. It is the Super Mecha that reanimate David, of which the voice of the opening is a representative, "the Specialist," as Ben Kingsley's figure is named, presumably in what the ending stages as Monica's de-extinction. De-extinction – a reverse pre-inscription that at once apprehends that inducted into the screen, or wall, as extinct at the point it appears reanimated in the arteficed stream of recoded marks and mnemonic triggers. The Super Mecha appear as pure cinematic figures that now alter matter, download and replay memories, screen sets, conjure environments and stage de-extinction fêtes. Screens scrawl across their "faces" instantly, marked by neither race nor gender nor organs, which elongated finger digits relay on touch. One of the opening shot's implications, in double retrospect, is that "we" are similarly unaware of experiencing not animation but something like de-extinction in cinematic consciousness (which may be to say, so-called consciousness as such). *A.I.*, the odd title of the work becomes a proper name for its (first) speaker. It begins as two alphabetic initials then reiterated, or unpacked, in caps (... *Artificial Intelligence*), as if there were ever an un-artificial intelligence (uninscribed and programmed), as if there were ever an un-artificial mnemotechnic order. The opening screen posits a future-past that looks back on "climate chaos" as a given and bespeaks the rise of AI, as and from cinema, in conjunction with it. The sentience of the screen evokes what Hitchcock implies by "knowing too much" – a trope caricatured in the hologram appearance of the digitalized Dr. Know, returning to cartoon animation in hologram mode, visit in the film.

V

What occurs if cinema here regards itself and the communal hive mind it generates from projected inscriptions as the forerunner of autonomous or cognitive AI – indeed, from the cave walls onward? Cinema, instead of practicing animation as we assumed, all along instituted a sort of extinction/de-extinction effect. AI may be termed the only technically correct "Anthropocene" cinema since it speaks from after our erasure. The term "Anthropocene" implies not only speaking from after Anthropos's own extinction but some eye arriving to confirm (and admire) the ruins – a replacement species, say, an alien visitor, or here, subversively, the Super Mecha. It, the Anthropocene imaginary, implies a projected Hegelian recognition-to-come that would legitimize, read, honor its ruins or disappearance, and read it as geological sediment. But the film *A.I.*, while staging this (the Super Mecha), does so on behalf of any screen, any film that we have ever contracted with, here decoupled from and turned as if toward us (no longer in the frame). Moreover, since this practice of extinction/de-extinction applies to cinematics from cave walls to the totalization of digital screens today, usurping public and private space, the screen also implies that the "Anthropocene" never took place as other than a "fairy tale"? The fairy tale is that there would be a future witness

and confirmer to the extinction; that is not what extinctions do – it is what filmloops defer and market. Thus, *A.I.*, the film, tracks a first robot's emergence to autonomous or sovereign consciousness, a "fairy-tale" quest triggered after the boybot David is first "imprinted" by his organic "mother," only to be then abandoned on the roadside like an unwanted pet.

VI

Throughout, *A.I.* identifies the robot generations with allegories of cinema – the bot Sheila's decoupled "face" from the machinal grid underlying it captures the dilemma of every screen animation, the mirage of face, the situation of the director, or viewer: what the opening voice discreetly totalizes. The Flesh Fair's director-barker and the bot dismemberments for the audience mimic demand for screen violence and torn bodies; the visit to Dr. Know; Gigolo Joe, whose prostitution mimes Hollywood. The Super Mecha outline even precedes David's first appearance, inflecting that it pre-inhabits all of the screen figurations and replays. Spielberg contributes two tropes to cinematic theorizations in the era of climate extinctions: the identification of AI with cinema as its precursor and the replacement of earlier tropes of animation and spectrality with de-extinction. As if to mark: it was over from the beginning to the extent that animation occurs over this decoupled redoubling in advance, and any "Anthropocene" a chapter, series, or phantom within its arc. "Cinema," in this sense, goes back to the cave walls, where marks and lines put in motion by torches coalesced the hive-mind effect of shared inscriptions, perceptual programming, the ruse of mimesis, animation in the contest of the hunt. Before any alphabeticism or pictographies, the specters on the wall or

screen would invariably be the first extincted (megafauna), without "our" portraitures. Holocene Park.

VII

When David returns to Dr. Hobby's clinic in inundated Manhattan, he is looking to be made "real" (a "real boy") so his "mommy" will take him back, but encounters his identical double, surrounded by books and reading. He beheads him in a rage ("I am unique!"), only to then find a David-factory disclosing him as not even a copy of a copy. Desolate, he pitches into the sea – where, after two thousand years, the Super Mecha arrive and defrost and r-animate him: he, David, is an "original" from the time of the humans, now extinct (whatever caused the nuclear winter). What *A.I.* conjures is a cinema without us, pure cinema. David's quest to become "real" masks cinema's own will here – to become, to control, "the real." If any "Anthropocene" can only be confirmed after it is long gone, to appear as a geological mark, it must be read by another set of eyes after that disappearance – or seem to be, perpetually, by cinema's archival survival. Gigolo Joe complains to David that they (Orga) hate "us" (Mecha) because after they are gone, "we" alone will be left, while "the Specialist" tells the analog and human-appearing David that he is the archive of the human period. The Super Mecha represent what Derrida, who avoided cinema, nonetheless called the "medium": "the absolute simulacrum of absolute survival" – wherein the first "absolute" must be heard to annul or withdraw the second as sheer phantom ("absolute survival"). Thus, for Derrida, "the future belongs to ghosts."

VIII

What, though, does it mean to have said that *A.I.* is at once the only literally "Anthropocene" film and that it discloses there never was an "Anthropocene" as other than an episode or serial program? The fairy tale in the film-loop would be that any eyes arrive to grant recognition and continuity. Again, that is not what extinction does. *A.I.* conjures in the Super Mecha a pure cinema without people – without mouths or eyes, skin color, apparent gender, pure cin-animation. Regarding us in the opening shot, it chooses, as David does with Monica, not to tell us, but rather keep up the anaesthetizing fairy tale. There is a reason we witness today's shift beyond tipping points and reversibility, paralyzed, as if we were watching a movie, spellbound. At this point, the voice bifurcates and assumes the lethal position of corporate Hollywood, the guarantor and agent at once of the "Anthropocene" snuff film – entrancing mass consciousness with extinction narratives, with the fairy tale of the Anthropocene.

IX

A.I. ups the stakes by citing Hitchcock's *Vertigo* by way of musical motif in the scene where David is abandoned like an unwanted pet. One may place a marker here, as *A.I.* goes hyper, and Spielberg makes his move to supersede *Vertigo*, or update it digitally – at which point the film passes from a network of doubling and inescapable circles to a vortex. It might seem a suicidal or hubristic move (as everyone knows, in the Anglo canon, no one overleaps *Vertigo*). The closest thing we have for a signature of cinema, of its backlooping mnemotechnic dilemma, is the vertigo-swirl or biomorphic coil emerging from the

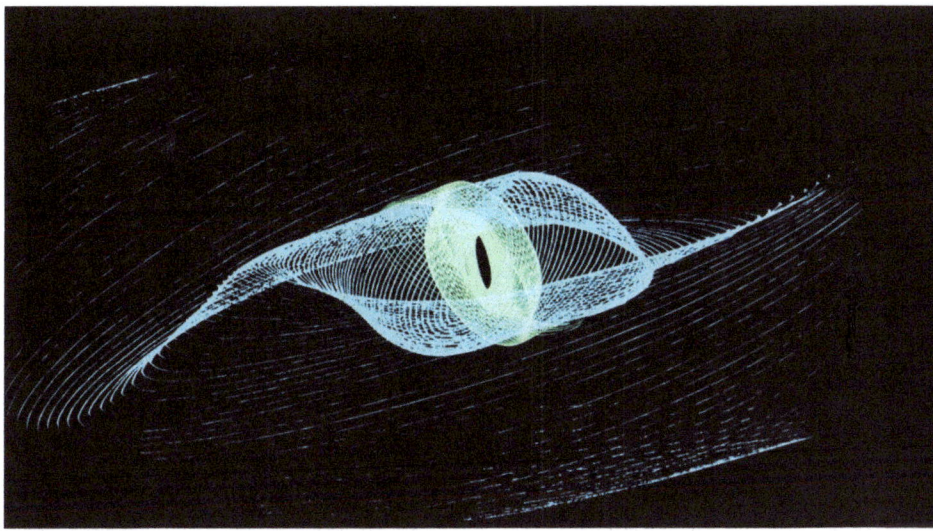

blond woman's eye in the film's credit sequence – expanding from that graphic backlooping of a mnemonic coil, itself implanted, which then exceeds the eye and, exported into cosmic formations, mimes the black hole of a galactic eye woven in Möbius bands. If the vertigo swirl is a signature for "cinematic" sentience, projected, edited, implanted, a circuit in which "the past" (as if there were a "the") would be inscribed to generate a "present" that cannot escape the accelerating filmloop of artifice, it recalls the logics of the vortex today and the superstorms as we pass beyond "tipping points" and initiate a cascading acceleration on auto-pilot. One might read this cinematic vortex was exported into the biosphere itself, and back, a bipolar vortex "today." Spielberg's pop contribution to the theoretical vocabulary of cinema is not about spectrality or mourning but its inescapable practice of extinction/de-extinction – what he rehearsed on his CGI dinosaurs in *Jurassic Park*. If that were the experience by digital consciousness of itself, there is no wonder that we witness ourselves, today, passing irreversible tipping points of mass extinction events as if it were all a movie, decoupled, fascinated. "Cinema" was not the recorder and archivist of the Anthropocene, as is suggested of the still analog-like David himself, but both agency and guarantor of the extinction that the latter hoped would give it definition.

X

Missing from the account, of course, is "climate panic" or whatever you want to call this trap and aporia, which only a robot could love, and only a Hyper Mecha aping "Ben Kingsley" – pulled from "our" memory banks in order to win us over, we de-extincted and pre-extincted – would appear to mourn. Some cinematic habits die hard.

This vortex for us, we old-model anthropoids, is a rotating precession of memory now simulated and captured by digital troll farms and stoked as perpetual reactivism, vortex-like, and perpetually too late. It exploits resentment as a structure of "consciousness" after mnemonics has been turned into tele-advertising and populist memes. One can see why the gentle "human" voice faked by *A.I.*'s Mecha-narrator is dubious – much as the Specialist's sit-down with David bears the whiff of pedophile grooming. We see the Hyper Mecha communicate voicelessly and instantaneously by downloading data, at contact, their "faces" flashing screens. This is where, at once, what appeared an entropic circular logic tips into a maelstrom from the opening shots and "voice." "We," the de-extincted, are being patronized, to say the least, fed the sickening wallop of intra-uterine bliss staged for David in Spielberg's acidic add-on and end. Like the scenes with de-extincted Monica, for a day, we are coddled and put into distracted bliss by products by which Hollywood preps us to embrace retirement, if only out of cognitive exhaustion. It is not accidental this conclusion appears to stage an analogic form or mise en scène, in David's conjured home as contrived by the Hyper Mecha. It is as if all along our mimetic cravings and identifications, such as cinema contrived and withheld at once, may as well be completely indulged now, much as Trump says of methane release or accelerating the already baked-in, "worst" warming outcome: it is too late, so curl back into the screen.[65]

XI

But there is a consequence of Spielberg's reverse arche-cinematic meltdown (or -up), and that lies behind the splitting away of face and the poisoning of identification and mimeticism as Hyper Mecha

mind tricks such as any screen performs. It is routinely remarked how "uncanny" the effect of Haley Joel Osment's ("David's") face is. This just by soliciting blank protectiveness and being a magnet for disowned projections – particularly when mimicking his adoptive parents' ways – neither human nor not, too innocent and too alien packaged in the most tender, little boy for us. This is garish and unnerving when he mimics laughing, aggressively redoubled and without object, at dinner. But the work accomplishes what might be called the "invacuation of the uncanny" as a trope at all, since the POV is that of AI, hyper-conscious machines after the human extinction is long over: there never was an "uncanny" since it is we who played *fort/da* rituals, who pretended the *oikos* or home was a given or interior rather than an artifice. It is, has always been sheer exteriority such as the Mecha assume, having become the regime of "life." What is of interest, is that *A.I.*'s implicitly scalding critique of a backlooped entrapment in the tropologies of identification, "face" and recovered loss is indissociable from what drives the extinction in advance, the carrot held before the horse of human consumption, futurity, and so on. So: how does *A.I.*, of all works, see itself as a bizarre iteration of *Vertigo* – whose Madeleine is already an irreal bot, of sorts?

XII

The vertigo-swirl is the ultimate signature of cinema's backlooping inversion of the mnemonic production of realities and, not just Möbius-like entrapment, but complete emptying out, "loss," or shattering divestment, and the precursor of climate vortices once fully exteriorized or exported – as currently – into the biomorphic orders of the life-form regimes it accelerates into. The images of today's superstorms miming the quieter vortices of caving ice sheets and collapsing coral reefs, and the like, conjure the moment when the cinematic funnel, the vertigo coil and space-stretching nausea and loss of direction and fall, appear visibly exported (and the reverse) into the inorganic accords of the biosphere. The film work, animated AI, accomplishes the invacuation of the "uncanny" as a trope at all, since the POV is throughout that of AI, hyper-conscious machines after the human extinction is long over. Once again: there never was an "uncanny valley" since it is we who played *fort/da* rituals, who pretended the *oikos* or home was a given interior rather than an artifice. It is, has always been, sheer exteriority such as the Mecha assume, the regime of technic "life." *A.I.*'s vertigo funnel, the work itself, takes from us any fantasy that this soft-voiced Hyper Mecha will be there to sift, honor, read and mourn us – it is worse than CGI; it might be an insult that we need be told "fairy tales" in a voice drawn from our affective memories to be lulled into a de-extinct trance. To be, in Hitchcock's choice term, spellbound. If *A.I.* is about reading its own survival, not that of Anthropos, it would be in quest of a cinema without screen that passes into the real – is made the real, like David becoming a "real boy" – and which, in fact, is the entire world the screen and its faux narrator offer us. But as a sort of Pre-Cog itself, it would be pleased today. This vortex exports the accelerating apparatus of cinematization from appearing to inhabit humans and memory into the digital noosphere, and now into the biosphere and inorganic orders. That technicexteriorized itself from the virtual screens of cave walls through becoming machine through the digital control of the noosphere and into the biospheric and climactic orders. Or does the fortuitous passage over tipping points and the accelerating and hollowing feedback loops mime the inorganic's digital totalization? The current migration into screens and digital "life" appears inseparable from the deterioration of the received organic ecosystem. It is the opening shots of the ocean that accompany the voice which fuses rather than separates out or mutually eclipses the pair of AI and climate extinction, a frame with no humans, no mimetic figures aside from a CGI "voice" projected from a Mecha being without need for one.

XIII

> We're suffering for the mistakes they made because when the end comes, all that will be left is us. That's why they hate us.
> —Gigolo Joe

Yet the Super Mecha serve another purpose. When David encounters his replicant double and goes into a fratricidal rage, beheading him, the latter "David" is interrupted reading. He invites the first "David" to join him reading. The first "David" asks: is this the place they "make you real" – that is, not just turn the bot into a "real boy" but where cin-animation passes into the real, its screen dissolved or infinitely multiplied and universally installed? The second "David" responds, smiling, with a witty, if obscure, pivot: "this is the place they make you read." Then he is smashed, beheaded, and "David" proceeds into the hell of the David-factory and off the deep end, or at least a ledge. At the navel of this factory and turning point in mnemonic and biotic orders, there is some sort of reading going on, or was, that must be interrupted by the first "David"'s blinding rage to discover he is not unique at all, having been "imprinted" otherwise. There is a quiver over a place of cinematic reading allied to the navel or foreclosure here – the backfold of the vertiginous swirl or ribbon. Presumably, the protocols of AI reside there to be altered or advanced. "David" number two is precisely not the psychotic mimophile with Scottie-like delusions, but generous, sophisticated, unbothered. One might weld this eclipsed opening or rupture, the emphasis on reading (as the film itself re-enacts: word codes to "imprint" David, books and storytelling, Pinocchio totemized, Yeats repeatedly cited and presented as text by Dr. Know. One culls a certain crystallization of the AI reader to come, who is here the "fairy tale" of the Anthropocene.

XIV

The projection of transformative AI is this double narrative of cognitive and biosynthetic transformation and planet trashing, ecocidal tipping points, mass extinctions – and implies a species split now underway. One cannot separate the promise of AI of post-mortal techno-homos from the simultaneous passing of biocidal tipping points which, as it happens, right about "now" (this very instant?), appear to have shifted into the slow roil of cascading backloops, irreversibly unfolding. This shift from a gathering entropy, in which intervention seemed possible, to a self-advancing vortex on automatic acceleration, seems a subtext of the ICPP-ratcheted-up reports starting around 2018, accompanied by the cacophony of superstorms and megadroughts, Arctic meltoffs and so on (long list). There are many mutations in how we regard or engage temporalities from that point on, which "we" have not adapted to yet – or are kept from in the perpetual re-activism its arrival guarantees. (This is seen, today, in the manipulation of re-sentimentalities as the meme for digital tribalisms and the survivalist wars they portend.) Such a double-layered narrative – the conquest of and by AI and the passing of so-called tipping points triggering cascading feedback loops locking in mass extinctionmerges smoothly, the one against the other, even as they are decoupled, in discourse, as if kept in isolation or eclipsing the other. It is not accidental if they come together not just, if rarely, in cinema (*Ex Machina*, *Westworld*, delete reference to climate chaos), but *as cinematization*. .) This complete split of purported "reals" is only consolidated with the advent of homo Smartphone – in which the hominem organism farms out memory, information streams, *Sapien* status, entertainment and identity to a handy screen gadget and, more generally, an unending labyrinth of screens as such. It is no wonder that cinema has tracked all this in advance – as if the era of cinema had all along been the advance guard of this inhabitation, easing us into screen universes, machinally crafting memory, softening us up for abdication (today's IQs are reported dropping with Western sperm count).[66]

XV

If all of what is deemed intelligence is "artificial" strictly, even when embodied in and as a niche life form, the artefaction has no one maker. In our purview, or that of the screen recounting its origin myths, its derivation from memory technologies and marking systems yielding the organization of imprints and projections was cinematic and became industrial cinema and detached screens, implants, and *AR*, and so on, today. The Hyper Mechas looking back on "David" as a twentieth- and twenty-first-century cinematic archive of the human event, Hitchcock appears already to have been there for Spielberg. *A.I.* ups the stakes when Spielberg both marks and is absorbed by this encounter with

Hyper Mecha precursor "Hitchcock" – yet another machine within the split-apart face. Spielberg invokes Hitchcock (as, more outrageously, does Tarantino) as a certain thinking of cinema by itself. Three times Hitchcock is overtly cited and dialogued with by *A.I.*, each attempts at appropriation. Each splits apart the "face" of the screen's fiction or voice or "fairy tale":

- A first invokes the psychotic-seeming amnesiac in *Spellbound*, Gregory Peck, in a murderous cinematic trance, approaching the sleeping woman-mother with a razor cited weirdly when the boybot David hovers with scissors over the sleeping Monica, in order to take some hair – tricked by Henry to magically earn her love (which hair later in fact is used to de-extinct her for a day). The blank memory of Gregory Peck's zero-likened figure in the post-war moment is marked by the void bearer of a missing "new head" (as the film refers to his rotating arrival at the sanitorium Green Manors): here, any personality or identity, like Peck's appearing as the murdered doctor, is artificed over the empty stare of the screen, and this is channeled to the bot-boy craving reciprocal love from the imprinted bonding on his non-organic "mother" figure.

- A second is even more vertiginous, which is a direct transformation of a core scene on cinema in *The 39 Steps*, the machinal character called "Mr. Memory," bearer of a formula for mass destruction, a silent (cinematic) bomber turned against the homeland, presented as vaudeville in the opening scene: he only records "facts," things of the

past, like the claim of still shots, entertaining a raucous, lower-class variety show (a cinematic enclave for which Mr. Memory substitutes for the screen). Spielberg doesn't expect anyone to recognize this, and it is a dialog within and between cinematic logics, but it is directly mimed by the visit to the CGI hologram of "Dr. Know" in the raucous Mecha red-light zone, traversed with Catholic statues of the Virgin Mary, a bespectacled "Dr. Know" crossed with a Yosemite Sam's cartoon moustache. The two Mecha, Gigolo Joe and "David," now question Dr. Know, who is paid like a dancing chicken generating fortune-cookie sayings – but in algorithms – and he first dispenses with "flat facts," then to yield to a hybrid with "fairy tales" so "David" can allegorize the generated answers, setting him further on his regressive quest, as a bot, to experience womblike encirclement or pre-uterine bliss. Mr. Memory for the CGI phase, when "flat facts" of analog media are pre-inhabited by a now shared, or totalized, AI artifice still, like the film, mockingly endowed with knowledge of the past (of the Hyper Mechas' own genealogy), which the entirety is narrated from the perspective of to our de-extincted selves.

XVI

The third invocation is, however, abyssal or vertiginous, if you like, and washes out the first two. It involves and invokes *Vertigo*. And it occurs early, before the mayhem of "David"'s dangerous quest begins, in the richly green forest scene in which he is abandoned and Monica

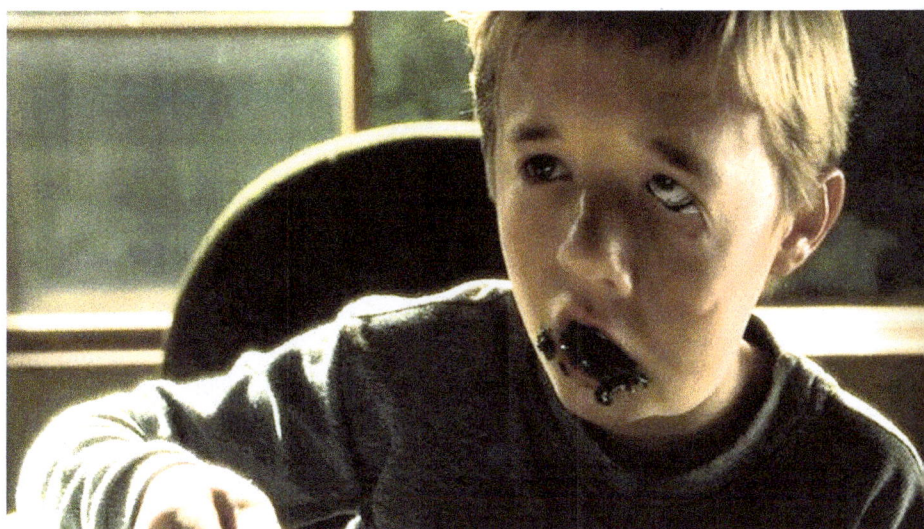

tells him to run off, like a dog rejected by the family left by the road. He wails at his (non) mommy not to abandon him – and, of all things, there are musical citations of the *Vertigo* score as Monica runs to the car (again, by a bona fide Hitchcock collaborator), yet the scene itself combines two, impossibly: the Muir Woods scene among the giant Redwoods, the oldest living things ("David" is not able to die, obversely), and the ending bell-tower scene of Scottie's total devastation as Judy/Madeleine goes over the edge for real, and the vertiginous backloop closes out an irreparably broken-limbed Scottie, a shattered or voided puppet. What on Earth, so to speak, does *A.I.* have in mind – that is, the logic that puppets "Spielberg" into self-erasure, here, of his contest with Kubrick, of his sentimental signatures, of his recuperative post-modernism – putting the invocation of *Vertigo*'s shattered closing of its circle with Madeleine's fall and the Redwood scene, jointly, with "David" being dragged through it all with a faux Orga "mom" abandoning him? "David," a non-sexed cyber-boy "imprinted" (words, reading and texts dominate the film's underbelly) with his non-mom? It is like "David"'s face after eating green spinach, melting to the side!

But Spielberg's attempt to override Hitchcock misfires: it is, after all, not the circularity of the narrative, disclosing a patronizing "Masterpiece Theater" voice speaking after our own extinction and, in Monica, temporary de-extinction both, that stands out, but that such circularity or seeming entropy in advance, with cinema's advent, tips into a vertiginous fall, an accelerant feedback-loop, which, however, is the formal condition of mnemotechnic humans or their programmed ticks (mimeticism, identification, face). Behind the entirety of *A.I. – Artificial Intelligence* is an attempt again to recast, and one-up, *Vertigo*. Why?

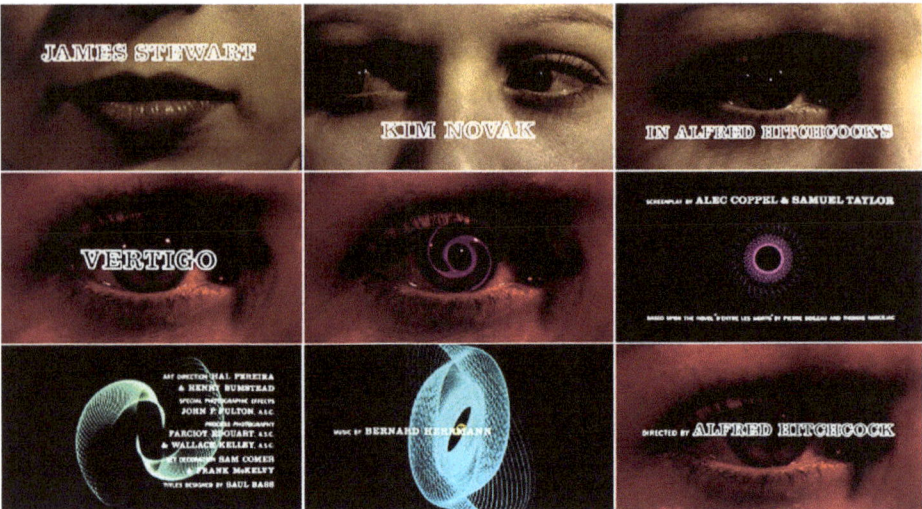

XVII

The unreadable biomorphic-swirl emerges first in the blond woman's eye opening the credit sequence of *Vertigo* and expands into a biomnemonic graphic coil both in and preceding the paranoid eye itself, Judy's or Madeleine's or another's (we are too close to identify a face). It is inescapable, irreversible, derives its "presents" from effacing backloops, implanted or artificed – rendering Dr. Hobby in the position of Elster. Perhaps that is why I'm drawn back to it – or perhaps, it is because of what has emerged to view as we observe the logics, on Earth, of its accelerations. These correlate not only with massive climate disruptions, and not only a weird identification in recent cinema with the justness of human extinction. Thus, Naremore's question cited above, accounting for his emotional grief: "The film [*A.I. – Artificial Intelligence*] is about the robotic post-human, and it uses a technique [CGI] that's occasionally described as 'post-cinematic.' Am I weeping for the death of David's mother, for the death of humans, or for the death of movies?" When a vast entropic order tips into backlooping acceleration, it is, resembles and spawns vortexes. Like the fairy tale that there is a good "Anthropocene" or escape from irreversible climate chaos.

XVIII

If *A.I.* is not simply Spielberg's work, it is not because he overwrites Kubrick's project – diverting it, adding an at-first-panned, Spielbergian ending, and so on. Rather, it is because the fully autonomous AI that is generating Ben Kingsley's disembodied voice for its audience, we later learn, has usurped the work's logic from the faux-fratricidal star directors, as if the work should either be existential-Freudianist (Kubrick)

or adolescent-, if formalist-, recuperative (Spielberg). Instead, the voice that speaks from or as the screen and who speaks literally of climate chaos and the rise of AI (as events, however, of the then past), does so, unseen I will say, as the title of the work, which we might hear as a proper name – odd as it is, two initials, unpacked as redundant verbal monicker. "A.I." is the screen before one, which speaks for the Super Mecha who evolve after the human extinction from the immature bots in the story. Of these David is the analog breakout model leaping into cognitive autonomy. With this opening frame "cinema," detached from us, speaks back as the screen from after our extinction – and, in citing the ocean as origin of animation, *before* that intervening event The speaking disembodies and encompassing screen, which we later understand is generated by Super Mecha, identifies something like pure cinema with the historial arc which yielded autonomous AI and digital totalization – shedding the intermediary Orgas through whom animation evolved. Gigolo Joe will tell David that humans hate them, Mecha, because, after the former are gone, only they will remain, the movie archive of what the "Anthropocene" was. But this is why *A.I.* may be the only Anthropocene film we have (except for all other films read through it) precisely because it also exposes the "Anthropocene" as never having existed, or other than as a "fairy tale."

XIX

Kubrick would have left the boybot, sunk in New York harbor, "conscious," perpetually, a copy of a copy, auto-abandoned, etc., but Spielberg has the Super Mecha arrive, who control matter, and haul him back, reanimate him, download all memory instantly, and then in reverse de-extinct "Monica" to spend a day of carefully directed pre-uterine immersion and *Liebestod* before erasing her from leaving any trace of having existed in time-space archives. But the trap closes as the "Spielberg" ending turns against itself as *A.I.* proceeds beyond the various Oedipal or Cain and Abel evocations that sputter or are set aside, indifferently, as is David's faux fratricide of his double, only to enter what turns out to be a "David" factory and to peer through a hanging skin of his own face. Face is what the Davids sells, yet it is renounced and broken apart by the screen, like Sheila's detaching Mecha face in Dr. Hobby's seminar or the Super Mecha – without face or sense organs (eyes, mouth) other than pure screens. The Super Mecha flash-replay the entire "film" for themselves from David's downloaded memory, to which we are invited as de-extincted viewers (not addressees). It is what Gigolo Joe means when noting that they (the Mecha, but also the cinematic archive) are hated because when the humans are gone, they will be all that remains. From the cave walls to the algorithm-driven screens, "cinema" had been the locus of a hive mind. Since the term "Anthropocene" contains in itself its own erasure – that is, it can't name itself as a geological strata that is only evident to eyes long after its departure or extinction, like any other geological epoch, the phrase assumes the position of other eyes-to-come, aliens arriving from elsewhere (as the Super Mecha first trope for the viewer), who will grant the Hegelian recognition that is built into the impudence of a self-naming term, Ozymandias-like: since it can per definition *only* be geologically named with a backglance, after it is long gone, it projects this position in so naming itself, simulating the missing recognition required from an other, in this case an other species or sentience form (ostensibly). (A.I. is in its way a short-cut or cheat sheet to this confirmation, one that for now even co-habitates.) But precisely this "Anthropocene" is the *fairy tale* the filmloop provides we long and perpetually now de-extincted viewers, no longer in the screen. *A.I.* not only cancels the "Anthropocene" it alone can conjure. The screen speaks from after our extinction yet from perpetual anteriority. It cancels said "Anthropocene" in a second way. Identifying autonomous AI with cinematic apparatuses, or "cinema" as A.I.'s stalking horse, reconfiguring human video-sentience, *that* would have always been the case – going back to cave walls, the coalescence of archival marks as networked figures of the hunt, mobility, "light," animation.... Any "Anthropocene" would be another moment or HBO series within *its* passage through today.

A.I. paradoxically reveals the "Anthropocene" as a cinematic construct, a fairy tale we tell ourselves. The vertigo swirl that began in Hitchcock's work has expanded beyond the screen, manifest in the accelerating feedback loops of climate change and technological advancement. Are we living the Anthropocene, or simply playing our part in cinema's grandest and most consequential de-production? Perhaps, *A.I.* suggests, the distinction no longer matters – in a zone where artificial intelligence, climate chaos and the algo-cinematic appear inextricable.

Hijacking Media – a "'Theoretical' Documentary"?

Hypothesis: That cinemacide occurs with the displacement of the "biopolitical" as fossil trope in the self-cancelling hyperbolics of the "theoretical documentary" – in which the media-enclosure of the hijacked bus as telepolis is situated techno-geologically.

10 Exploding Omnibus – or, "This Ain't No Fucking Movie!" (on *Bus 174*)

> After the Candelaria massacre, do Nascimento used to sleep underneath a bridge near where he hijacked the bus. It's right next to the Globo TV office.
> —José Padilha, interview[67]

> We were putting on an act. It was an act.
> —Janaína, *Bus 174*

Today's occasion raises the question of a "new real" of cinema – Brazilian and Argentinian first, to be sure, but also more broadly what *cinema* does with the "new reals" of twenty-first-century horizons.

These are both clearly mutating today, both cinema and today's post-global horizons, and both are inextricable certainly. The era of hyper-industrial modernity can be called a "cinematic era," even as it careens into a dawning era of climate chaos – resource depletion, extinction events, food riots, and so on. The latter can be a name for material processes before what lies outside of its bubble of reference. And if one wanted a single metaphor for this anthropic bubble, you might think of us as all on a *bus* – its windows are like screens that look out, in safety,

or can be seen into, its transport appears stationary, and its wheels imply a circular backloop that puts time itself in question.

I will speak about this bus. And I will do so by relating José Padilha's *Omnibus 174* to a very distant and unlikely parallel, Hitchcock. What links these is the use of a bus as a figure of cinematic experience itself. *Bus 174*, in its Anglo title, presents itself as a faux documentary deploying archival footage of a real event – the famous hijacking of the Rio public bus by a "street kid," Sandro Nascimiento, which brought Brazilian media to a standstill. As such, it is as far as possible, we might suppose, from high modernist allegory of the sort Hitchcock practiced, wherein the enigma of cinematic processes are ceaselessly foregrounded, marked, put in question. One can fail to notice that the film is traversed by figures of media. In Padilha's "Rio," the real has been already cinematized, the product of the encompassing Globo network of Brazil, of telemarketing, and telenovelas, of video-screens that penetrate into the *favelas*. This association of the "street kid" Sandro with media (as in the interview cited above) is emphasized, of course, in the film: Sandro contrasts his "act" with that of action movies and is quoted by his adoptive mother as performing the hijacking to become a screen star, of sorts, even if he would not be there to then watch the screen with her. This screen, like the bus, will itself encompass that which is literally before the viewers of *Bus 174*.

For Hitchcock, a bus is recurrently used as a figure of cinematic experience (as is, more frequently, a train). In Hitchcock's early film *Sabotage*, this takes the form of bombing such a bus itself. A boy unwittingly carries a time-bomb onto a bus together with film canisters. This temporal bomb blows both him and the bus up on the way to London's Piccadilly Circus (called "the center of the world").[68] What Hitchcock implies is a technical arrest of time turned against those who imagine themselves transported by this agency, communally protected – rather the way Hollywood, at the time, would. In Padilha, the bus is different. The bus, which is also the screen or film, takes into itself the community or telepolis ("Rio"). This entire circuitry will be hijacked, turned on, put at risk or held in hiatus. [69] What occurs when cinema turns against a "real" it has produced too successfully, to the point of producing a suicidal double of itself?

The interface I invoke between Hitchcock and Padilha's bus is meant to disrupt a literal reading of the film as a social documentary and exposé of human injustice. Were the latter the case, one of the film's aims would be to humanize the hijacker Sandro and retrieve his narrative, thus returning him to an object of mourning or recognition.

The subject of *Bus 174* is media itself, which Sandro is associated with and takes over in the hijacking. The omnibus of the title first appears on the screen – after an extended aerial panorama descends into the city – as a clip of public footage from a recorded traffic cam. It is thus at once the *real* bus (cam footage) and cites the automated surveillance cam. Various forms of media and its devices are thus incessantly *cited* in the various frames and sequences: handsets, walkie-talkies, televisions, towers, writing on windows, video-game screens, Sandro's references to movies he's seen, and so on. If the point of the film were restorative, to return to us the personhood or identity of the "street kid" Sandro, as the dirge-like score suggests, that would mean we have absorbed him into the "visibility" side, that of the humans and the community. In a perverse way, this may turn out to be so when the film gives the "real" Sandro fame and then generates another film in which a "real" *actor* is playing and interpreting him (*Ultima Paraida 174*) – that is, when the cultural machine reduces him to a psychology, arching his disruption. To restore Sandro to "one of us" would be to restore the control over visibility that was interrupted, shattered, hijacked, upended by this takeover of media. Instead, *Bus 174* attempts, like Sandro, an intervention in the whole at the sacrifice of itself as a "movie."

I (In)visibility in the Telepolis

What does cinema do when it encounters a "real" that has already become cinematized in advance? This is particularly so if it encounters in that world itself what has come to be thought of as a biopolitical closure.⁷⁰ A bus, after all, is an enclosure, and it has an outside. What would cinema do if it had to turn against itself, against this bus, to

These boys battle against invisibility.

break the spell of that real, which the film calls by default the "visible" as such? Would cinema be acting all but suicidally? If such were the case, if the "real" already were produced by media circuits, and if this in its way generated and defined the visible, then the visible would be blind – and cinema would have to reorient itself. It might erupt, like Sandro Nascimento. In *Bus 174*, the "street kids" are repeatedly identified by the enlightened sociologist (Luis Eduardo) as "invisible," as bound by "invisibility." For example, we hear: "this Sandro is an example of the invisible street kids," "these kids battle against invisibility," "these boys move about the street invisible," and the screen literalizes this by blocking out faces, covering with scarfs, and so on – with Sandro doing both repeatedly: putting his face before the camera, addressing "Brazil," asking to be filmed, then wrapping his face again and again. A persistent defacement suffuses the film's seeming parade of speaking faces.

The "street kids" are repeatedly called non-visible and non-beings, "garbage," "bare life" in Agamben's much-used sense. In human form, they are excluded from human recognition – a ban that constitutes the orders of the human or "visible" in turn. It is a global situation, today, corresponding not to a class split but a species split. The "street kid" is not a lower class but an out-of-life (as *bios*). Žižek represents this view when developing Agamben's example of the slum-dwellers and *favelas* to "the camps": "the "slum-dweller is a *homo sacer*, the systemically generated 'living dead; or 'animal' of global capitalism... pushed into the space of the out-of-control."[71] For those who find the homo sacer trope a misfire from Roman Law that tames and situates "bare life," the parade of the defaced or faceless echoes with being, all along, outside of the bus, ex-framing its wheeled and windowed *enclosure*.

Yet it may be this biopolitical analysis is precisely what is undone, in a way, by *Bus 174*, or exposed as itself a movie we reel again and again because it maintains a nostalgia for the *bios*, the idea of a polis even without polity that requires a binary of the "excluded" and the "invisible" to fuel itself. Sandro, the totally excluded, the category of "bare existence" (*zoe*) in media revolt, appears all too included even at the margins within the circuitry of the telepolis. In question here: what occurs when the "invisible" erupts to take over the screen of the "visible," hold it hostage in real time (35 million TV viewers, "hundreds of eyes"), or when the homo sacer is in media revolt? Similarly, what occurs when media is itself held hostage, as if itself arrested? Padilha replicates Sandro's violent takeover, and does so by an attack not on the inhumanity of the citizens (the sociologist's moralism), but on a real that has already been cinematically programmed to include, over and over again, the fragile hypothesis of the excluded.

On the one hand, *everything* in this film is edited footage of the "real," real TV footage of the event, real witnesses, real place-shots – no authorial intervention at all. On the other hand, everything arrives as a citation, from the cuts of footage of the event to the gallery of talking heads and voiceovers.[72]

It turns out, then – this will be my hypothesis, minor as it is – that if *Bus 174* brings things to a halt, holds media hostage like Sandro, it is in part because it refuses to be a movie. Like Sandro, when he tells the cops again and again, "this is not a fucking movie." It does not aim to disclose some unseen truth as would what Padilha dismisses as "observational documentary." Everyone knows the "street kids" are the unliving, the undead, the "invisible," they are openly visible at the same time, and so on. And it does not aim to recover the human subjectivity of Sandro for us to understand him. Sandro will wait for *Ultima Paraida 174* (2008) for that to be tried, for an interpretive narrative to psychologize him or make him like us, and to end up in Wikipedia.[73] Both would be symptoms of relapse. This is made clear by the second tracking shot of the mountains where we find them crowned with giant media towers – overseeing Rio, feeding screens and circuits.

Mere *documentary* is to Padilha as historicism might be to Benjamin, a deceived index for producing the real (or visible), at odds with the citational hiatus of the image, the "enemy" (Benjamin says).[74] What occurs in the "theoretical documentary," though, is that rather than recover or mourn a real Sandro, the latter is produced increasingly as a counter-logic and figural agency. For the sociologist, who romanticizes this, Sandro becomes the one who sacrifices himself to enter the orders of the "visible" and burn out in "fiery glory" – a vaguely Christian mytheme: "the boy exchanges his future, his life, his soul, for an ephemeral and fiery moment of glory... a crucial moment. A turning point." One might ask, instead, at what point does he appear as a certain master-of-media and a saboteur of the cinematic "real"?

What do I mean by this?

II A Documentary of "Theory"?

Padilha cites a clipping from the film itself in the commentary he addends to the DVD for the English audience. He suggests, speaking a sort of halted global "English," that the film is not a documentary but something that he calls a "theoretical documentary." The word "theoretical" is cryptic and points to what could be called the allegorical zone of which the clip is example. It is opposed to what is called the "observational documentary," which investigates a "contemporary" event. By contrast, the former takes up a "past event" (in this

case, what is also already a media event, one which leaves traces in the archive as what he calls the "most filmed" hijacking ever). The "theoretical documentary" then destroys the representational premise of documentary as its premise. The phrase is less a self-canceling oxymoron than a non-existent or nameless genre, of which it is the sole example.

Padilha's commentary is interwoven with clips of Sandro provoking the police from a bus window. It captures Padilha giving his account of what he is doing while having Sandro direct and speak for him. Other things are marked in this performative explanation of the non-existent *"theoretical* documentary" – for instance, the identification of the bus with cinema itself with its succession of screen-like windows and the writing on the window glass. Padilha marks his own identification with (and eroticization of) Sandro: he, the director, will in fact take directions from Sandro about when to get "on and off" the bus when editing – that is, when to go from the event footage to various places and interviews explaining the episodes of Sandro's existence. Sandro will in fact act precisely as the director of a film he is improvising for real inside and outside the bus itself. He tells his various hostages how to act for the camera, in effect creating two different discourses (in the bus, and outside, for the police and beyond them, the cameras). The hostages, moreover, will be told to act as what they nonetheless are, that is, to act terrorized, while Sandro assures them they will not be shot – staging such an imaginary shooting for the police. Padilha will note that the director's most difficult task is to take the "right shot," also a problem for Sandro and for the hapless cop, Marcelo – who rushes him when he finally leaves the bus on a seeming whim, drawing the drama to an arbitrary close – who shoots from a "foot" away, missing Sandro yet managing to shoot the hostage in the face (triggering Sandro's shot as they go down). Padilha identifies with Sandro's takeover of the cinematic bus, the "visible," as the film itself takes over the name of the event itself (*Bus 174*).

This pre-reflexivity gets worse once one begins to read it, spins back on "itself." When Sandro tells the cops again and again that this, whatever is taking place, is "no fucking movie," it occurs in what Padilha nonetheless calls one of Sandro's "speeches." Here is Sandro do Nascimento – who, today, has a feature film made after him, with a real actor acting him. And he is telling the cops, now from the screen, that this is "no fucking movie." At the time of saying this, "real" time, he is already performing as if in some film (he has models in his head, like the airplane movie he refers to having watched the night before). And, of course, he told his adoptive mother, Dona Elza, that she would see him become "world famous" on the screen, even if he would not be there to watch it himself. And Sandro speaks, here, in a movie of sorts in which his name is listed among the players.[75] Nonetheless, to say this is "not a fucking movie" speaks, for Padilha, of his *Bus 174* itself. To say that a "theoretical documentary" is "not a fucking movie" is to claim it cannot be digested as a film is, gives up the ostensible fictions of cinema, and can only be regarded – like Sandro's suicidal trajectory – as tearing a hole in the confabulation of the "real" in the media circuitry that is called "Brazil" (and, again, indexed to Globo itself).

When we see Sandro talking to the camera and the cops, we think we know what it means. Sandro acts for and is protected by the cameras. The governor intervenes to assure that he will not be shot by a sniper, since brains splattered before the kids watching the TV at home would be bad for re-election. By saying that this is "not a fucking movie," Sandro is telling the cops that he really, really, will use his gun and massacre people on the bus. He is giving the cops a way to envision the disaster to come (blood everywhere) and insisting they take him seriously. This is "serious shit." Yet it produces the opposite effect, since just as his threats are an act, as if in a movie (he does not intend to kill as he says), so the cops act even more as if they are watching one. "This is not a movie" means he will kill for real, really for real. He will act. Yet your rhetorical position is weak if you have to insist to the cops they are not just watching a movie – and, indeed, the cops seem to just pull back and watch, doing nothing, as does Brazil at the time on TV.

But – and I will soon give this up – what is meant now when the screen itself speaks through Sandro, as Padilha does? What does it mean when *Bus 174* says, to us, this is "not a fucking movie"? That it is not a documentary, not a narrative film? Then what is it? To say "this is not a fucking movie" is to ask where the performance is an intervention in the real real ("real shit") – not that it represents for us a real problem (the life of the "street kids") but that it intervenes at the archival site from which "visibility" and the "real" are as if produced. For *Bus 174* to say this of itself, that it is "not a fucking movie," would be like cinema suiciding itself to break its own spell. And one can see why, since the film appears as if it were a composite of archival footage and sheer reportage. There is no narrative imposed, not fictions, no Padilha, reportage and monologues of witnesses. That is, it seems to insist on the unadulterated real of all of its edited clips. Nothing seems added (except for the imposing score). Yet everything acts as a citation rather than a report.

When Sandro tells the cops that this is not an "action" movie, then, he implies two things: he acts like the villain of the action movie, which the police must be "phoney heroes" in; yet he also implies that he is the hero inversely, rebelling against a totalized system that deprives him of life and exterminates his kind (his "little friends" at Candelaria). Turning the tables, Sandro criminalizes the entire order of the "visible" which the police enforce and the cameras represent.

Which is why the clip is chosen. If "documentary" is allied to the indexing of a real or praxis (the supposed other of "theory"), then what is called "theoretical documentary" turns on a "past event" and alters the event's condition of emergence. That is, it is an act that alters both the past and, prospectively, the future past. It does this not by restoring a lost identity or narrative (Sandro as person) but by allowing what is called "Sandro" to take over the bus, violate the orders of the "visible," criminalize and expose those orders. When the "invisible" takes over the cinematic bus, "Sandro" in effect opens an irresolvable rift within the visible that cannot be dissolved, reversed or explained in the latter's terms. He wants to enter it, become a media star momentarily, become visible or recognized at the cost of life, but also de[r it. In this instance, it is the theoretical dimension that makes some sort of irreversible or sabotaging action possible – as other, say, than a mere replay of facts.

The term "theoretical" adheres to the Greek *theorein* – echoing seeing and theatrical spectation. When we think of this term as connected to abstract thought – as when theory is opposed to praxis – we ignore that it has to do with the eye or where the latter is bound to memory and media. It returns us to media, or in terms of the film the telepolis itself, what is here called "Rio." But there is more. "Theory" implies here what Benjamin might term "allegory," an act, the active intervention into the archival site from which the "real" itself is generated.

If *Bus 174* were about recovering Sandro's otherness and humanness for ourselves, the living, that would be reassuring and an end of mourning. But then, it would be documentary or perhaps the feature film that retold the story yet again with actors and with a moral interpretation, *Ultima Parada 174*. It would be a mere movie. But *Bus 174* instead creates of "Sandro" a monstrous rift between the visible and the invisible. He becomes an ante-figure that resists readability absolutely. Instead of being returned to us, to a psychological or moral story, he drags the order of the visible – which is also the screen – into this rift, or across to its other side. The latter occurs when the screen goes negative, exchanging black for white, on the third prison visit. Starting off as the criminal, the "bad guy" of the show, Sandro criminalizes "the visible" as such (which includes the viewers). This frees what we call "Sandro" from any final resting point or grave (no one attends the funeral) and makes him most real, so to speak, when he has no place. The "theoretical documentary" is "not a fucking movie." It produces something monstrous, is something monstrous, a hybrid, twenty-first-century non-genre of which it is the only example, a post-biopolitical cinema (if there were ever other). This is why *Bus 174* seems in endless mourning from the beginning aerial shot, with its dirge-like score, and at the same time beyond mourning. But the object of mourning is not Sandro but "Rio," which is to say the bus, which is also to say the mediatric "polis."

III Exploding Transport

It is here that I interrupt the "interrupted robbery" that is *Bus 174*. The bus as a figure of cinema is a not unfamiliar "modernist" trope and runs, like that of the train itself, through Hitchcock. Again: the series of windows like screens, the faux transparency of the glass, the stationary transport promised to the one seated, the wheels that elicit a backlooping logic of time and artificial memory circuits yet also provides a material base. In the opening of *The Man Who Knew Too Much*, this occurs as James Stewart and Doris Day banter about the desert world outside the windows – everything is first made "familiar," compared to what is known (America), looking through windows of utter alienness. The windows as screens place the *tourist* viewer in this position, but the America referred back to as familiar never quite existed as such. As the "little boy" Hank says – who hangs between every question and says "I don't know" a lot – what they call the "dark continent" is twice as bright as their home city, "Indianapolis, Indiana." "Indianapolis" is the name of a phantom mediapolis of sorts, called "home." This cinematic bus at once too dark and too bright

reminds the viewer that there is no "America," which was originary to the Indians. As Hitchcock remarked: there are no Americans, since they are all foreigners. But this cinematic logic is casual compared to that in *Sabotage*, where the saboteur – who runs a movie house – places a time-bomb on a bus that is blown up. The bomb of time is placed next to movie reels, hence cinematic, and under a birdcage in which it is timed to go off for when the birds' sing. The saboteur is linked to Hitchcock and the cinematic bomb, meant to get at some real, to act and alter the very "center of the world" (as is said), must explode the bus of media itself – a suicidal gesture. There is something of this gesture at work inversely as Sandro perpetually counts down to a putative or proleptic "six" o'clock which will, as needed, appear plastic or put off or given up.[76]

If the figure of the *bus* is viral, determining its own point of disappearance, it is because, in a sense, it is a figure for what can have no proper figure, just a stand-in, since it is encompassing for those inside. Hitchcock would explode the bus of cinema by its own means, while for Padilha, the "invisible" hijacks the bus and the "visible" itself. They converge about the attacks on the construct of the "visible" itself. Padilha turns against "movies" insofar as they motor and perpetuate the state of the real, the telepolis. If Hitchcock's figure in the 1930s anticipates the tele-empire which Hollywood will be indissociable from the spread of, Padilha arrives at the world as that has taken place. It is a world that cannot be held in place by the biopolitical model that, like the sociologist, can map the cruel management of the "invisible" – the separation of the polis or "life" (the citizens) from the waste, the "invisible," the excluded, the nonhuman (the street kids, the animal). What one witnesses in Sandro's uprising and takeover is not the revenge of the *homo sacer* – as both Agamben and Žižek cite the *favela* to supposedly demonstrate. Rather, what is called "Sandro" represents an uncloseable rift in the orders of the "visible," into which gap the entire bus is drawn and arrested by. It won't close, because the very tissue of the visible/invisible distinction is breached and "Sandro" as the name for a performative within the film cannot be mourned (no one but Dona Elza attends the funeral), despite the ritual of recovery the film acts as if it is performing. Instead, something else occurs, since in standing outside the bus, which now includes "Brazil," the camera attaches itself to something outside of the biopolitical map. It is not so much a zoopolitical cinema – that is, shifting to the side of the nonhuman (Sandro as "street kid"), or the thing or animal – as one that is post-binarized altogether. The telepolis is overseen by the media towers, and its public space is not the street but the screen itself or the stream of Globo data.

To be reactivated from his "real" deed and transferred as Padilha's media warrior, Sandro must not be restored to a personality and history (which only leads to a feature film explaining and erasing his uprising). To be "real" in the theoretical documentary, one must become a figure that defaces the circuitry, that hijacks the bus itself. For Padilha, one can follow where this leads and why, finally, this turns against cinema itself. What Sandro triggers in *Bus 174* is not only a rewriting of a past or the partial realization of it as an event (Sandro would have disappeared without this film). What is little noticed is that there is a virtual meltdown through media and the archival supports of the telepolis – one that leads as if to a non-site, the dead space of the third jail visit and the graffiti on the walls of the "Vault" in the second visit.

IV An "Interrupted" Interruption

It is the second appearance of the mountains above Rio that displays the strange media towers, at once assertively dominant and fragile in the lush peaks. The row of towers appears as a blind panopticon struck into the earth and ranging over the polis below – managing media circuits, feeding TV screens, relaying memory streams. It is blind not because it emits rather than sees, creating the "real" that pacifies so surveillance is irrelevant: it is blind since, rather than maintaining the "human" polis, the entirety drifts, suicidally, toward twenty-first-century mutations that put its survival in question. When, throughout the

Sandro made some friends in here.

film, different faces turn up masked or blurred, it is also before this tower in a way.[77] Sandro will both thrust his face out the window and wrap it, giving and taking away face. In the biopolitical accord that is put on display, the visible itself is invisibly broadcast, generated.

Let me follow this thread. The circuits of media that traverse *Bus 174* are established from the first – with the camera's aerial panorama. It is free of gravity and independent of the footage it will soon cite, including each talking head it visits. The arrival of footage appears in identifying the bus – a traffic-cam shot in the street. It is a robotic camera, a surveillance wired to no one. It links a new public space to the camera stream that has no locus and whose secret is, in part, the death or arrest of politics as we imagine it – the ghosting of the polis. The polis, or "Rio," has become a transindividual memory circuit as on the screen's parade of talking heads that serially supplant one another yet, collectively, traverse Brazilian types and roles (students, police, social workers, intellectuals, street kids). So many faces, highly individual, and yet always battling against a stripping of face itself – the blurred or covered faces of interviewed gangsters, street kids, police.

Sandro arrives with films in his head, raising and lowering his face scarf in ways that make no sense. He is in the remote tradition of a Quixote or Bovary in that regard – transposing a pop-media template into life, into action, and because it is cinematic, it is condensed to an instantaneous occasion. He refers to the airplane movie he saw the night before as what he would not do (he could not toss a hostage from the window of a bus) and had prophesied to Dona Elza she would see him on the screen even if he may not be able to. Sandro is protected so long as he holds the cameras hostage, become hostage to them in turn. He leaves the bus suddenly, surprising everyone, as if he tired suddenly of the eternal recurrence he finds himself.

But while the film poses as a faux documentary, the foregrounding of media had marked itself from the start: cell phones go off, cameras snap. Rather than strip cinema of anything but sheer reportage and archival footage (cited, edited, in dialog with Sandro), *Bus 174* discreetly runs through a series of mock televisual or cinematic genres like so many windows behind which a black figure with a gun moves. Padilha cinematically signals and disowns these genres in turn: there is the action film, the CSI-like TV police show (jazzy cuts and musical score), the horror movie, the docudrama, the black comedy, the *favela* thriller. When the cop thinks Sandro was shot and is shocked to find him reanimated, the unkillable monster movie flares up. And it finally displays the real "real" of a snuff film in which Sandro stars. When he is asphyxiated in the back of the police van, the cameras, which revoke protection once he leaves the bus suddenly, turn on him and go into a feeding frenzy. They prevent themselves from seeing through the windows by their own reflected lights on the glass. This runs through the commentators too: the therapist references "American films," Sandro's performance is called a "show" (seen at home on TV), and Sandro's erasure in the van is said by a reflective cop to fulfill the demands of the script, since in any show "the bad guy has to die."

The *window* is, itself, a hero here: framing, admitting visibility, capturing, reflecting, itself as if invisible and yet a barrier. A screen that reflects another scene, it can draw attention to itself materially, become a text. A shift occurs – still linked to the media tower – when Janaína must write script in giant letters backwards on the glass. This writing is done with lipstick, eroticized, and shifts attention to what lies in between, medial, the glass. The transparency of the glass becomes opaque and replaced by script, the media now become a mute barrier. This appearance of giant, reverse letters hosts a rebus. It draws us into a mute archive, a meltdown through technologies of script preceding speech or talking heads. This trajectory once again ups the ante for the violation of "visibility" Sandro performs. It again redefines the reference of the bus, and draws us to the underworld of inscriptions.

Such focus on what we are calling here "archives" had all along been at work – footage, testimonials, public records, places visited by Sandro revisited, scoured for traces.

Bus 174 descends from the sun to the catacombs and vaults. It turns out, it was or is all about archives – and its proper title, *Omnibus 174*, marks this confluence of the bus with a totality that is poised by the "omni." Thus, a certain muteness and muting opens beneath the many voiceovers and monologues. Cops don't have walkie-talkies and make hand signals like mimes; Damiana's stroke makes her mute, so her daughter has to read her diary for her. We pass through a series of public archives, typed jail reports on Sandro, court sentences consigning his place in time for the record. And this drifts toward what is called the "Vault," the dead inner sanctum of Sandro's prison cell, without sunlight and now windowless. The "Vault" provokes horror in the prisoners, who panic once they hear that they are assigned there. Yet the total enclosure of the cell is like a bus without windows but instead with walls covered now with unreadable graffiti. The "Vault" inverts the open sky of a panoramic earth outside *Rio* and precedes what is called visibility. What goes on in this underworld, this darkroom?

Unlike the third jail visit, this one is emptied when visited. The warden stutters, saying it is "not a jail," unable to give any name to the placeless place. Here the unliving banished from the *bios* are stored outside of time and life, warehoused. In turning to the cell wall, again, the camera finds the parallel to Janaína's writing on the window in the very minimal of inscriptions on the wall – graffiti from unknown authors, unreadable name-words: *Grota... C.V... Deus... Orpheu* (lower right). The wall has effectively sealed over the window on which letters had appeared that could be read still. It becomes the extreme reduction of the bus regarded now as enclosure even as it turns to mere inscriptions. Only when the cinematic bus is apprehended as what the film calls "Rio" is it clear that that, too, is encased by the walls of the "visible." It is blindly constructed, suicidally when viewed from outside. The order of inscriptions in this lightless zone gives rise to sight. The name "Orpheu" appears, what does it mean? Is Sandro a different black Orpheus in Brazilian cinema? Does Padilha mark the lens itself as Orpheus-like since, in turning back (the "theoretical documentary"), it must annihilate and lose the beloved again, leave the mourned Sandro to the underworld? Does one find, at this omphalous of magical inscriptions, an abandonment of mourning – from an increasingly non-anthropic position?

The network of images, cams, devices, windows with writing, archival files, recorded exchanges, interweave as a narrative generating (and abrupting) mode of A.I. in which the screen curates, yet which is in hiatus before the camera descends and arrives, as in a closed loop or confinement that runs from the open sea and sky, framed and deployed, to the sunless etchings on the jail pen's wall.

But the camera does not stop there. Rather, it asserts and shows itself at times, like Sandro covering and uncovering his face. Thus, when the aerial panorama that takes us over the mountains identifies with a terrestrial surface, "bare life" of sorts without face (which is to also say personification), the camera eschews any anthropic position. The camera enters from outside. It departs, moreover, from the watery surface of the sea outside Rio, not just the encircling element out of which life fashions itself but the specular surface out of which reflective visibility and the "eye" is as if created by a techno-genesis of "light" outside any human cognitive design. The descent from above might appear discreetly angelic at first as it discloses the *favelas* creeping up the cliffside. But it is suspended between heaven and earth, birdlike. It citationally invokes less a Christian or angelic visitation to the unjust world of men than the repetitions and decays of cinematized history and empire: the descent into Rome and the Colosseum in *Gladiator* or that of the *Triumph of the Will*, with attendant resonances. At the point at which the hapless cop Marcelo rushes Sandro to earn his promotion and, as an expression of Brazil's inability to sort out any perspective or target, shoots the female hostage Geisa in the face instead, which we all want to understand, the screen goes into slow motion, frame by frame, invoking its technological powers to taunt the eye with what it cannot even then see.

V Sunless Vault of Inverse Inscriptions

The screen itself can also mutate in this general dismemberment. The tear in the system that "Sandro" holds open circulates virally. The screen can mark itself within the frame or become like the window overwritten with reverse spellings. Upon the entry to Nova Holanda slum, where Sandro spent time, the entire screen is taken over by a video game on which cartoon creatures blast away with superguns. It is placed on the way into the *favela* – adjacent to the "STOP TIME" motel sign – to mark that this circuitry of media in fact precedes, and informs, the *favela*, that video-game shoot-outs become the template of gang gunplay (a cinematized "life"). The video game is the brute interactive programming of *cin-animation* shaping the definition of life and the order of *favela* violence – machine-gun posses performing video-game battles. It again undermines the biopolitical map since those radically excluded from human status are not strictly "excluded" at all. They are clearly within the mutating circuitry. One is now within a suicidal biopolitical accord running on inertia, in which living film-loop, the "visible," is entirely aware of the "invisible," sees it daily, and acts as if it were not there – accelerating its. Sandro's apositionality stand's in not just for the fable of the non-person, street kid, nationless, or sub-caste but, in today's amplifications, the occlusion of climate chaos that this totalized bubble, "El Globo," cancels from visibility (in the Trumpocene literally banning reference or acknowledgement, the science and the green sprouts jointly. This model of *telepolitical* circuitry coincides with a suicidal acceleration and blind, which corresponds to its virtual blindness to what is outside the bus, the way global financialization is blind to the economy of extinctions of climate change. And the sociologist suggests that there is no way out of this, no amelioration or political change since the system pursues an "efficiency" (biopower) that does not exist. This is why to speak of a post-biopolitical cinema is redundant as a formula. And it is why one is not tempted to speak of Sandro as Agamben's *homo sacer* in revolt, or of this even as a zoopolitical cinema, since this would keep the trace of its binary in place. Here the definition of the "excluded" would have to be revised, since nothing lies outside this circuit in fact: "Sandro" can end up in Wikipedia and have a film made after him, and the *favelas* have become a film-making colony of sorts, with German boutique hotels and tour groups.

Sandro commands the bus, waving a gun, exploding from invisibility, giving "speeches," a Hamlet of the *favelas* and of the screen. He hijacks the bus on the absurd pretexts of robbing students, becomes "Sergio," puts the bus in hiatus, expends himself as a caesura. He has an accord with the director's cut. Since what is implied by visibility includes how the human community establishes its identity as that, blindly, there is in fact here no human that is defined by this accord as normative – not the "street kid" banned from human life, but also not the social order that persists as if it had a biopolitical contract or was even efficient. Many references are made to "Brazil" as being just the opposite, foremost in this the police (as demonstrated by Marcelo's misfire when a foot away).

I suggested that the opening aerial panorama bonds the camera with the non-anthropic – the reflective sea surface it emerges over and rushes across, the mountain cliffs that rear about Rio, then the media towers. The first gives rise to the visible and to specular "life," the second gives place to the polis, and the towers sustain and traverse perception and memory. This makes a certain "Sandro" also a ghost of the future, of the human or polis itself trapped in a suicidal posture with reference to what is outside the bus – the twenty-first-century horizons mentioned above (biodiversity collapse, resource depletion, extinctions). If hyperindustrial modernity had been the era of cinema in all of its extensions, the bus could denote the Anthropocene era, "Rio" the post-global tele-necropolis from the camera's perspective. This bus is interrogated and turned against by the camera, which has the power to place the screen itself in the third prison visit, in which the negatived order of the visible appears to pass to its inverse side in its entirety (as if white for black, black for white).

The drugs that Sandro is hopped up with (cocaine is mentioned) are not cause for his dismissal. These course through and militarize the *favelas*, are tied to their sovereignty within the mediapolis. And they mime the anaesthetization or drug of the "visible." The reason that this cinema is post-biopolitical is because "life" as an effect arrives in advance as cin-animation. When the film descends for its third visit to its underworld, to the now anonymous site called "any jail in Rio," we see what this movement to the other side of the screen of the "visible" entails, which is still the screen. Nothing is excluded from this circuitry, no outside, since it itself has become pure exteriority.

The media's presence makes the hijacker confident.

In short, cinema as the viral bus – brought to a halt, arrested, held hostage – destroys the biopolitical binary that is a residue of twentieth-century humanism in hedged, inverse form: it does not view, record, trace, burn through, deface, from any point of identification with a nostalgic external premise or fold (*zoe*) to which the telepolis is segregable. This Padilha more or less tells us by identifying Foucault as the intellectual guiding the middle-class, social-worker students, tourists in the *favela*, whom the director delights in burning, shooting, and humiliating in *Tropa Elite*. As Justin Read describes, this new megalopolis is one of circuitry, a "unicity" that mimes the screen as self-deconstructing site of a viral public memory (and mourning): "The Unicity would be a break with both biopolitics and abstract space in certain respects. In the space of the Unicity both the natural and the symbolic (spiritual) have been liquidated by the sheer physicality of networks.... There are real modes of disaffection, disempowerment, and exploitation in the Unicity, but none of these has anything to do with binary relations of inclusion/exclusion."[78] In *Bus 174*, the cinematic trace identifies with the broadcast towers and auto-cams, the seeming meltdown through media forms, and the sea surface and primal mountains cradling the hero of the film, "Rio." Sandro appears, in this sense, as an unreadable, black figure that is identified with an explosive cinematic logic in faux revolt. Padilha puts on and off the face mask of documentary as Sandro draws up his perpetually slipping face cover.

VI Cinemacide

Having entered the film in the open air without men, in the reflective sun on the watery surface or atop giant terrestrial folds, we are taken for the third time underground and in a sunless enclosure. The scene is more than a living tomb of warehoused human lives placed outside of time – held, like and unlike those in the bus, in a non-place. Thus, upon entering this one scene, the screen itself appears in shock, flipping the interface of the visible outright. It retreats into the form of the filmic negative, white exchanged for black, entering the unearthly Hades of Žižek's "living dead," the storage space of "bare life." Unearthly faces and bars. Nonhuman forms swarm, rush up monstrously behind irrevocable bars. The bars seem to graphically form their own serial interruption. If the "human" constructed itself by banishing from the "visible" its waste as bare life (to pursue this formula), then the former would be here shown to be inhuman all along. The previous empty jail cell called the "Vault," which Sandro had spent time in, by contrast mimed, in its rectangular walls scribbled over, the old trope for the isolating frame of the celluloid band. In this case, the negatived visit to the sunless crypt also mirrors and inscribes the screen's citizen viewer.

Though stationary, arrested, the hijacked bus of media goes hyper, has a citational meltdown that leads to the third prison. It is, or goes, "out of its mind" (as is said of Sandro at the end). Erupting into visibililty, speaking from different windows on the bus, Sandro is protected by the cameras. He is not a killer, we are told by the therapist and Dona Elza. Padilha calls his behavior "poignant," his harangues "speeches." He fakes a killing by shooting the floor (off camera) and tells the hostages how to "act" terrorized before the camera – how to act as what they "are," so that he does not have to *actually* act (shoot them). Everything takes place in hiatus, in putting the moment off, in counting down to six o'clock, then extending the countdown – the "STOP TIME" motel sign. Sandro proceeds by directing his not-a-movie to approach "serious shit," the virtual collapse of the "visible" regime. This involves different discourses for those inside and those without, who can see only threats, savage terror, gunshots. Janaína accounts: "Parallel conversations were going on. One for the cameras and public and another for the hostages inside." She and others talk of the confusion, when the lines of play-acting and lethal act crisscross or converge – with even Sandro not knowing which is which: "Then he looked at me and he seemed to realize that there was something ambiguous in what he was doing. Either he wanted us to put on an act or he really wanted to kill us." Sandro occupies the position of a permanent parabasis or disjuncture that, rather than being recoverable, mourned, recognized or given stardom, cannot be closed but takes into its black hole everything in contact with it – or the bus. Stalking the interior, visiting the bus window as if a rampart, his "do not forget" is the Candelaria massacre that he witnessed, in which the police murdered his "little friends," and his mother's virtual beheading by robbers in front of him as a child. These two ground zeroes of catastrophic memory, before which all else is erased yet which are not locatable in time, make of the film's Sandro a zero figure, in the inverse sense to of a "man" to come.

This doubling of acting and the act returns us to the act that the "*theoretical* documentary" implies of altering the effaced or anterior event to shatter or transform a captured present – what remains if one is "not a fucking movie." Sandro imitates "himself," becoming his dark double, the name that the police give to him and which he accepts to play, "Sergio." There is a perpetual rift, here, in the always missed performance of a self in auto-citation. The film does not, in this sense, do what Padilha says or the documentary as social criticism pretends. It does not only retrace for us the lesson of how torture is institutionalized in the state, as Padilha encapsulates in his commentary. That is local, and everyone in Brazil seems aware of it, even to look the other way at Candelaria. The zombies of the third prison, negatived, are both the strangest beings of the screen and, discreetly, form a specular relation to the viewer, the citizen, the blind prisoners of visibility.

What the screen calls "Brazil" is in fact a place in which nothing signifies itself, nothing equals itself. The cops are not cops, cannot shoot, and drop stones on the heads of sleeping "street kids" (of which the population approves); the hijacker cannot hijack, has no clear aim or exit; the prisons are not prisons (as the warden says, unable to give it a name). Indeed, the prisons permit mass escapes on a ritual basis, as if recirculating "bare life" as needed. And documentaries are not documentaries. This is what a caged prisoner, appearing negatived on the screen, left to rot long after his time is up, implies in saying: "Nothing in Brazil works."

Thus, the act is what the "theoretical documentary" itself poses – and makes interminable – as it steps beyond "movies." It destroys documentary by turning its citations into an active archive – a circuit of which a certain Sandro emerges as the anarchic center to which all is related, or referenced. Such a work would precede where that "real" is itself generated, artificed.

This is the import of the camera's identification with the water's reflective surfaces and the mountains, with the physics of light and the nonhuman, whatever is outside the cinematized spell of the *polis* – or bus. The "theoretical documentary" aims to precede the condition of emergence of the event it would account for, and this by an act it can never coincide with.

Where the order of the "real" is produced by mediacratic circuits in advance, cinema, like the hopped-up Sandro, goes hyper, empties itself of itself, saturates itself with its doubles. It becomes monstrous and gives itself a counter-task: to interrupt that, to take over the bus (or itself), hold itself hostage. And in doing so it turns itself, suicides itself in order to sabotage a bad cinematic "real." It *acts out* cinemacide. Or in Hitchcock's terminology, it would explode the bus itself, which is where the phantom of cinema begins, absorbing and initiating the reign of *afterlives* that encircle the globe.

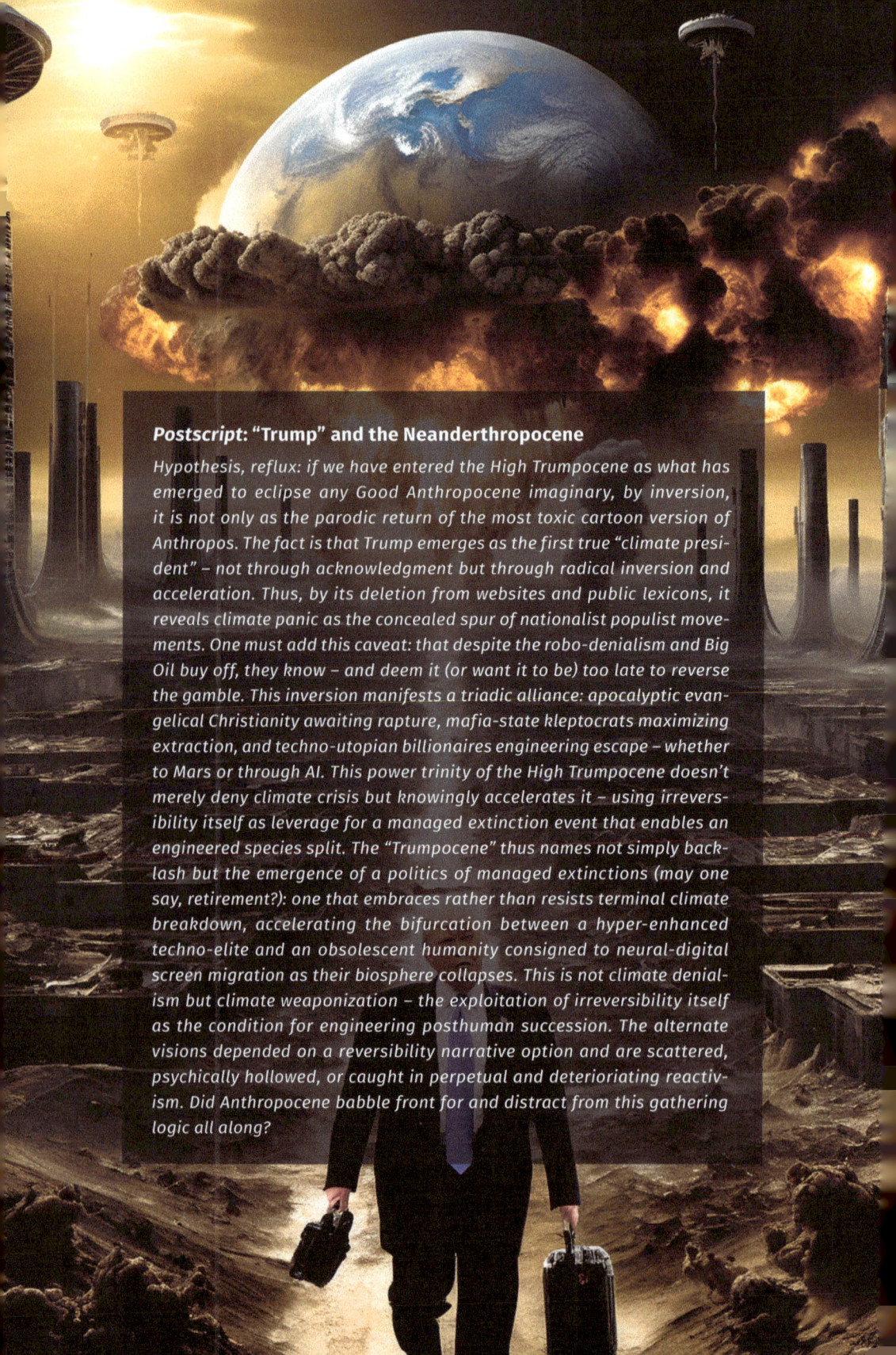

Postscript: "Trump" and the Neanderthropocene

Hypothesis, reflux: if we have entered the High Trumpocene as what has emerged to eclipse any Good Anthropocene imaginary, by inversion, it is not only as the parodic return of the most toxic cartoon version of Anthropos. The fact is that Trump emerges as the first true "climate president" – not through acknowledgment but through radical inversion and acceleration. Thus, by its deletion from websites and public lexicons, it reveals climate panic as the concealed spur of nationalist populist movements. One must add this caveat: that despite the robo-denialism and Big Oil buy off, they know – and deem it (or want it to be) too late to reverse the gamble. This inversion manifests a triadic alliance: apocalyptic evangelical Christianity awaiting rapture, mafia-state kleptocrats maximizing extraction, and techno-utopian billionaires engineering escape – whether to Mars or through AI. This power trinity of the High Trumpocene doesn't merely deny climate crisis but knowingly accelerates it – using irreversibility itself as leverage for a managed extinction event that enables an engineered species split. The "Trumpocene" thus names not simply backlash but the emergence of a politics of managed extinctions (may one say, retirement?): one that embraces rather than resists terminal climate breakdown, accelerating the bifurcation between a hyper-enhanced techno-elite and an obsolescent humanity consigned to neural-digital screen migration as their biosphere collapses. This is not climate denialism but climate weaponization – the exploitation of irreversibility itself as the condition for engineering posthuman succession. The alternate visions depended on a reversibility narrative option and are scattered, psychically hollowed, or caught in perpetual and deteriorating reactivism. Did Anthropocene babble front for and distract from this gathering logic all along?

11 Notes on the "High Trumpocene" – or, the Discreet Charm of Climate Autocracy in the Time of Cascade Events

…the Anthropocene epoch, from which it is a matter of escaping as quickly as possible…
—Bernard Stiegler

To think about the Anthropocene is to think about being able to do nothing about everything. No wonder the topic inspires compensatory fantasies that the solution lies in refining the bottom line or honing personal enlightenment – always, to be sure, in the name of some fictive "we"…
The trouble starts when this charismatic, all-encompassing idea of the Anthropocene becomes an all-purpose projection screen and amplifier for one's preferred version of "taking responsibility for the planet."
—Jedediah Purdy, "Anthropocene Fever"

I

Since the "Anthropocene" was introduced as a term by Paul Crutzman – ostensibly giving the present epoch a new proper name, the age of man – it has served as a magnet term for disciplinary and ethical probes, many remarkable, yet it has been caught in a certain vortex, a trap: not merely the mistaken allure of the proper name itself, but its function as cover for an engineered species bifurcation already underway. Never fully defined or "officially" claimed (that is being sorted out, but a bit literally), contested by alternative master terms that float by interestingly (a "Capitalocene" (Jason W. Moore), a "Chthulocene" (Haraway), and so on), we forget the term is primarily literary and designed to captivate and distract – precisely as it obscures the weaponization of climate crisis by those positioned to exploit its irreversibility. It announces both techno-mastery of all life, and its own disappearance or extinction at a stroke: no one can "say" this term since it requires eyes after its disappearance to confirm the geological stratification whose abject literalism is required to gain note: a drama, and con, of auto-inscription, a bit desperate. Yet this very desperation masks a more cynical calculation: the understanding that passing tipping points enables rather than prevents certain forms of mastery

– specifically, the acceleration of techno-eugenic transformation for a select few. If one had full mastery, said *anthropos* would not be facing logics of accelerating extinction – unless, of course, this extinction event itself becomes the lever for engineering posthuman succession. Thus, the term "Anthropocene" may have been already oversaturated and exhausted after a stellar, twenty-year run, victim of the celebrity culture it rode, but more crucially, it masks the very species bifurcation it enables through its ethical hand-wringing. More interestingly still, it may come to be regarded as a complicit distraction from the real climate opportunism taking shape in its shadow.

This is what Jedediah Purdy concludes in a review of its trajectory: it has accomplished nothing aside from "Anthropocene talk," and produced no unified "we" for the imperiled humanity the speech act implies – indeed, as the Trump counter "revolution" plays out, as nominal democracies pop and a recoil to artefacted tribalisms, walls, territorial markers, autocratic and neo-feudal plutocracies, something has shifted in the terms of writing or its videoscaped spin-offs. The shift is more radical than mere tribalism: Trump's presidency marks the emergence of climate-crisis governance through acceleration rather than mitigation. As it stands today, the proper name "Anthropocene" is caught between generated double narratives: that of a great techno-mutation of human life, enhanced and immunized with gene-editing and AI interfaces, mnemonic implants, gifted longevities, and so on, and that other, which predicts accelerated climate chaos, systemic collapse, resource wars (including the Arctic, Mars, and so on), and a "banality of the Anthropocene" (Stiegler). Yet these narratives are not truly opposed – rather, the acceleration of climate breakdown enables the techno-eugenic transformation, with Trump's triadic alliance of evangelical apocalypticism, extractive kleptocracy and techno-escapism serving as midwife to this birth of posthuman succession through managed extinction.

Clearly, the dubious proper name "Anthropocene" involves a claim to appropriate all hominid types and systems under the nominal Greek and Western abstraction (a covert cherry on the post-colonial and post-global stew). That is, it is the Aristotelian, or at least Homeric, *anthropos* conjured and enforced entirely on the gesture of exclusion (not not-Greek, not a child, not a woman, not a slave, and so on). Yet this exclusionary logic now operates at a higher level: not merely between human types but between legacy humans and their techno-eugenic successors. Nonetheless, just as the term "Anthropocene" traversed all major disciplines and began to congeal into an imaginary "studies" or knowledge network, Purdy observes it is also depleted and serves, in fact, to distract. It produced no "we" even as a "negative universal" faced with auto-extinction issues (Dipesh Chakrabarty's best stab). More crucially, it masked how climate crisis would be weaponized not as a call for unified human response but as leverage for engineering species bifurcation. It produced no effective response and no transformation – among scientists or critical thinkers, there would be a representational limit that is symptomatic of the trap itself. And it was on duty as tipping points emerged regarding the accelerating irreversible vortices, now, of biochemical and geomorphic mutations. It would be impossible to distinguish "Anthropocene talk" from being a Hamletian soliloquy complicit with the ecocidal acceleration.

It is not accidental, then, that Bernard Stiegler, reading the entirety of the human arc through the co-evolution of technics and mnemotechnics, casts the "Anthropocene" as a dead-end trap to be turned upon and "escaped" from immediately – an entropic swirl, against which he calls for a neganthropic and negentropic reset to keep open the prospect of future, transformed technologized communalities. Yet this very logic of escape has been co-opted by the triadic alliance of the Trumpocene: evangelical anticipation of rapture-escape, kleptocratic resource-hoarding for survival-escape and techno-utopian dreams of Mars-escape or AI-escape. The impediments exceed entropy itself. It is a distinction of the Trumpocene that active interests advance climate chaos with intent, leveraging it for kleptocratic orgies and *Walpurgisnachts* of a extractivism (the Arctic, seabeds, the moon, fully militarized). Purdy's humanist skepticism requires a democratic "Anthropocene" but the climate autocracies, inevitable in a one way street, had already arrived. Purdy's discourse does not allow that one can know and decide it is a waste to pretend some integral earth climate can be saved or managed, or even paid for. That a political power might see its caste interests for the future lie in accelerating the *entropic* itself – ripping up regulations, erasing climate initiative, accelerating emissions and coal use, banning the real of "climate change" as "fake news." This is not mere denialism but trolled denialisms for digital rubes to cover the opportunistic acceleration, which gets the Tech Bros all hot and bothered, who have already assumed the triage of the legacy Anthropos models as an acceptable price for the fusion of an A.G.I. godlet with organic hominids: Trump emerges as the first true climate president by weaponizing irreversibility itself.

What does it feel like to experience the Trump-regime takeover and performative inversion of rotted democratic legacies – with a cabinet chosen to "deconstruct the administrative state" through incompetence, sabotage and hostility in each case, a flurry of tweets, anticipatory counter-trolls and fake "fake news"? Whatever else, it is an approach that is clarifying, irreversible.

II

Matt Taibbi in his early caricature in "Trump the Destroyer" tracks the cross-eyed contamination, which is a psy-ops exhaustion that is perpetually unreadable as to its elusive goals:

> The genius of Trump has always been his knack for transforming everyone in his orbit into a reality-TV character.... Whatever your lowest common denominator is, Trump will bring it out and make sport of it.... The same phenomenon is now in play with the whole world. President Trump, following Bannon's lead, describes the press as an "opposition party" out to get him, and before long, they basically are.... We always assumed there was a goal behind it all: cattle cars, race war, autocracy. But those were last century's versions of tyranny. It would make perfect sense if modern America's contribution to the genre were far dumber. Trump in the White House may just be a monkey clutching history's biggest hand grenade. Yes, he's always one step ahead of us, and more dangerous than any smart person, and we can never for a minute take our eyes off him. But while we keep looking for his hidden agenda, it's our growing addiction to the spectacle of his car-wreck presidency that is the real threat. He is already making idiots and accomplices of us all, bringing out the worst in each of us, making us dumber just by watching. Even if Trump never learns to govern, after four years of this we will forget what civilization ever looked like – and it will be programming, not policy, that will have changed the world.[79]

Yet what Taibbi identifies as mindless destruction masks a more calculated acceleration: Trump's seemingly chaotic dismantling of environmental protections serves the interests of all three components of his base alliance – evangelical anticipation of end times, kleptocratic resource extraction and techno-elite preparation for species succession.

In a period of magical thinking, like our own, seancing the debris of the "Anthropocene" in search of an exit might compel extreme experiments, gnawing of trapped limbs, waiting for the guards to nod. Yet this very logic of escape has been co-opted by the triadic alliance: evangelical rapture-fantasy, oligarchic bunker-building and techno-utopian Mars colonization all represent perverted forms of exit strategy. I would ask, to break up such a seance, what would happen if we retired the term as ossified and crafted as a sandtrap, or had we never have taken it up with such an expenditure of innovation?

One might want to speak of the "Anthropocene" from the perspective of those arriving later, say, by mid-century or beyondafter this generation is vanished as a criminal memory, and Nemesis accelerates. The form that takes is abrupt climate riffs, cascade events, uninhabitability (AMOC, Arctic inversions, the most charismatic only). That is, one need think beyond our complicit and toxic generation that accelerated irreversible tipping points while negotiating their own safe zones. This very irreversibility is then leveraged and weaponized to engineer what some hail as posthuman succession ignoring that this will to cancel or exceed itself had always defined to "human." The "Anthropocene" might name, in retrospect, an unbearably culpable period of stunned chatter in which terms like "shock" and "this changes everything" would be masticated, while climate opportunists hastened species bifurcation. Surprising that none of this is mentioned regarding Trump – that they know he lies robotically but trust he does not "believe" in climate chaos.

Today would be the time window in which the possibility, if merely technical, of any evasion of the "worst" outcomes of global heating would be missed – an event, if not crime, of cosmic proportions, depleting the biotic options for a future comeback over aeons. Yet this very depletion becomes leverage for those positioned to exploit it. So, if the epoch of "Anthropocene talk" occurred as the vortex of bio-material acceleration took over irreversibly, that "epoch" of a couple of decades would be, in retrospect, inseparable from the open acceleration it, blinking, oversaw. Almost all of the discourse of the "Anthropocene," after all, has to do with how to maintain legacy concepts and political imaginaries in the face of climate chaos – from mass climate refugees that Europe pretends can be turned off as a matter of mere (climate) wars to the agonies of mass extinctions. What it misses entirely is how climate breakdown enables, rather than prevents, certain forms of mastery.

This is why the *High Trumpocene,* or "Anthropocene 2.0," rescinds the first "A" and its rationalist contracts as schemes to fix the game – the rhetoric of the magic pharmakon, golden-maned, of the post-truth gameboard, in which whoever trolls first takes advantage. Behind the recoil to what is marketed as tribal identitarianism is, after all, climate panic, on display everywhere (especially strategic robo-denialist firewalls). Or, overwhelmed by attention vaporizing streams, *a panic of reference* itself, one that correlates with the vast digital hollowing marked by A.I. pandemonium and Trump as resolute Nonreader and mnemonic implanter of alt "facts," the seizure and orchestrated submission of once bold grey matter (shrinking brain size, loss of critical skills and rise of post literacies).

It is not incidental that in this turn, we find, rather than the maintenance of the fiction of the "Paris" accords – that a fictional global effort to rein in climate catastrophe would be launched, the outcome as if in suspense for a century (the end of the century is the usual faux metric to get to, as if all goes dark beyond that) – that, in the case of Trump, blanket climate denialism is asserted together with the delegitimation of "facts," the production of post-truth media mills, in short, a bifurcation between self-enclosed, dueling media-microecologies and the material reals banned from discourse (much as the Governor of Florida, a climate-doomed topography, ordered the term climate change not to be used by state workers). Everything that "Anthropocene 1.0" had been speculating about arrives and flashes by in "Anthropocene 2.0" and entails something left uncommented: on the one hand, after "tipping points" pass, the politics of the earth becomes de facto a politics of managed extinctions going forward; on the other hand, this second phase "Anthropocene" dawns with a hyper-acceleration of AI, genetic engineering and designer babies, longevity boosts, mnemonic implants – that is, an entire mutation in what might have been called "the species" once, even though it conceals an engineered species split.

This convergence aligns with the massive wealth shift digitally engineered following the "2008 financial crisis," so called, which has left iconically six individuals with the equivalent resources of the lower half of the global population – a concentration of power that enables the engineering of species succession. The shift to the Trumpocene – short-circuited "facts" leading to closed media ecosystems, reasserting tribal ("human") identity before the anonymity of a globalist imaginary – which strove in the Paris accords, to shift to an "Anthropocene" perspective of a certain collective, a certain public good or commons, which was transparently bogus. Yet this very bogusness masks something more overawing: the acceleration and exploitation of climate irreversibility by Trump's triadic alliance.

III

So, the Trumpocene at once rips away the "Anthropocene 1.0" imaginary and, doubling back, forecloses the address of any real or "facts." The various declensions of *Alt* Right takes credit for circling back and absorbs the Alt Left – disenfranchised labor, claims on behalf of an artefacted "the people," aware that the left liberal imaginary had mangled itself, which in a sense could not even address climate change or climate refugees except in packaged tropes of "social justice" – or counter-imperial justice. The elusive invocation of a dike of climate chaos is not, however, social primarily, not between individuals and groups and so on, as it was not either for the Greeks – and introduces a nemesis born by and turned against the crime of the "Anthropocene" as such.

America inverts, preparatory to the decades of resource wars and sorting winners and losers, centers and triages peripheries in the decades to unfold. It is not incidental that Trump can be analyzed not as a hyper-narcissist media effect but a sort of algorithm, formed and accelerated by the memes and content that consolidate verbal power, entirely empty otherwise, a sheer technic or a foretaste (however obverse to expectation) of the "singularity." By this is meant that, rather than anticipating a biotech hybrid, the latter term in effect becomes an auto-memic algorithm without interiority – just as the rhetorical arbitrage would be a return and restitution of the tribal, the identitarian prop meme and reaction formation. Yet this very emptiness serves the interests of the triadic alliance: evangelical apocalypticism requires no content beyond end-times anticipation, kleptocratic extraction needs no justification beyond profit, and techno-escapism thrives on the very vacuum it helps create.

Thus, the Trumpocene is not the counter-revolution to the "globalism" that links up with "Anthropocene" responsibilities – the Paris accords – but an inverse parody and acceleration, the staged re-assertion of a certain *anthropos* sovereignty over chaos, inundation, contamination.

It is fitting that the "Anthropocene" in a sense pop and obvert, self-seal and accelerate, or be fashioned as such, just when raw tribal identities would be refabricated, the concept of a common *anthropos* suspended as genuinely irreal, a homogeneity of memes in closed and multiplying media ecosystems (and "safe zones"). Trolling itself, the introduction of an algorithmic "singularity," without content or identity, dangles alluringly to separate out the chaff. And all this in the face of the herded climate denialism, officially banished while doubled down on and actively, *intentionally,* accelerating. Of course, if the latter were already irreversible, and if the liberal order of the "globalist" regime now being rolled over were not transparently fraudulent to begin with, the logic would seem less ambiguous. Moreover, what appears as mere denialism masks a more calculated exploitation of climate crisis by the triadic alliance: evangelical anticipation of rapture, kleptocratic resource-hoarding and techno-utopian dreams of escape through AI or Mars colonization.

But there is a schism of sorts, again, as the traits of any "Anthropocene" mutate together with its globalist pretext. Parodied, inverted, instead of a unitary man with Enlightenment-gifted human rights, or even a failed "species" identity, we are hurled headlong back into neo-feudal mediacratic oligarchies – moreover, with a mocking exaggeration of mad Anthropos, yellow-painted mane, rapacious, extractionist, a Twitter fiend without attention span.

This is entirely transparent when *read* from the "perspective" of climate panic, even as the diagnoses of radical "inequality" miss the boat when wired to legacy narratives of capitalism, free markets, improved "equality," the fate of democracies, and so on: it needs to be seen entirely from the perspective of climate chaos as enabling condition for species bifurcation. This massive shift in terrestrial digital wealth under cover of the "2008 financial crisis" – when corporate media was streaming denialism in diverse ways – makes perfect sense when read against then climate predictions: the economy is engineered now not to suppress class struggle, or labor, increasingly irrelevant to an era of robotics and universal income (and last-man culture generalized). The massive shift involves Silicon Valley preppers – today's Trumpian Broligarchs – and prepares for who will control the resources, technologies, military environment and atmospherics decades hence. That is, the techno-mutant surviving caste, inbred progeny of the ".0001 %" and their support complexes. To repeat: the mass redistribution of wealth was already a knowing response to anticipatory predictions and knowledge. These were also quite apparent in the staged deferral (and comedy) of a "Paris," of tipping points passing short of overturning the economic gameboard.

If we take the "Anthropocene" as marking these twenty-plus years that encompassed both "Anthropocene talk" and a new super-elite perceived, or caricatured, as hinting global governance, then its overturn by the "Trumpocene" and the disarticulation of "Western" protocols, "values," legal premises, and so on, has to be read with suspicion – as the "post-fact" mediacratic circuits attempt to seal or enclose themselves (shutting out or banning climate chaos and what comes next). If you want to think of and with "tipping points" having passed, and a period of afterlife being projected (say, Stephen Hawking's two centuries or so to colonize other planets), one might illustrate with a favorite: the inescapable and sudden effect of mass methane release once the tundra is breached and thaws. Game over stuff. But this very "game over" scenario becomes leverage for the triadic alliance – evangelical anticipation of rapture, kleptocratic extraction until the last moment and techno-utopian preparation for species succession through AI or off-world escape.

A little googling behind the headlines reveals that is underway now, bringing scientists to tears but of no use even to the "mainstream" media or the fake news that tars the "media" itself as "fake news" and an "enemy of the people" (thanking old communist rhetoric for the lift). Here, the whirlwind of proactive memes from B framing A to destroy in advance reactive events begins an inevitable closing off, calculated, spun in an intra-cynical clustering of robo-identarian traps. The migration to the neural screen is complete, and bifurcated from the debris world. The era of climate chaos was never going to leave "democracies" intact, so this rollover would not be surprising, except for its Trumpian form in the case of "America," which can be read less as a white, nationalist, reactionary reflex than an attempt to assert strength, sovereignty and feudal autonomy against the import of ecocide, which it cannot effect or evade. Thus, the madness of Trump and the "vortex" he engenders and feeds off.

But the more obvious secret of the Trumpocene is that the giant factor he openly calls a "hoax" and to be dismissed – climate chaos (and the ethos of unleashed mega-extractivism just as the emergency bells resound at fever pitch) – had been his greatest ally, seeding the profound "referential panic," down to words and bytes and predatory *algos* pre-hacking and dismantling techno-mnemonics. *Earth turned against us, us*? Seasons upended? Cognitive speed outrun on all fronts? Migration into screen "ecosystems" of bot herded swarms? Daily modes of reaction formation and rage release channeled and churned? *He* makes it momentarily vanish, which is what "we" want. But the secret is he knows otherwise, as does his adopted son Elon, even when the latter in his X interview parrots that there is no problem with oil (except it may run out), and later that he bonds with "Jesus's teachings," to seal the triadic power: apocalyptic, "Evangelical," white, nationalist Christians, vengeful in plotting rapture; the looters and extractors on steroids (Trump) and the Mega-Tech Bros (plotting retirement of current human models and conquest of AI, then fleeing our doomed planet for Mars). Now, let's pretend that Trump and Musk know the climate advance is passing into cascade mode – and decide to accelerate (ah, that curious term) regardless, or even to leverage and monetize it at scale? Then the Trumpocene appears not as a backlash and revanchist swing but the politics of managed extinction flashing its twenty-first-century outline.

The "Trumpocene" inverts and cancels the "Anthropocene" – marked as a globalist logic, which climate chaos necessitates rhetorically, now supplanted by counter-elite "Billionare" cartels stripping the copper wirings – and arrives as its pop caricature (ruleless rapine of the hyper-extractor, Anthropos). And certainly the embarrassing mirage of a "good Anthropocene," which Musk toys with as bait (AI has an 80% chance of a utopian outcome in which everything is immediate and low cost for all – hmm; 20% that humans are erased). Yet, to return to my opening reference, the "Trumpocene" arrives also as the evil-twin double of Stiegler's neganthropic counter-strike. But this raises the question, again, of why the "Anthropocene" may not even be recognized as the *nom du jour* decades hence, be actively disowned and pinned on the current generations sorting out their dilemma – that is, inhabiting a doomed system that guarantees mass extinctions yet cannot be modified or abandoned in the necessary timeframe – why what seems, today, the Eurocentric vanity of a nice proper name, even the noble name of the Earth's master, might be eschewed and discarded as specifically referencing a time and population superseded, and better forgotten, the messy genetic soup of hominid modernities, become wasteful, extincting all in sight, not needed for labor, not evolved or enhanced or networked or gene-edited sufficiently for the next variant, as they think of themselves, primed to outlast the ecocidal descent – or at least play it out long enough to colonize other space rocks and extend the species game, or at least that of intellection and consciousness in non-organic forms.

Yet there is reason to mourn the demand to "escape" the trap or vortex, a figure which Stiegler deploys, and there is reason – a few years after his suicide if not before – to retire this rhetorical option. Stiegler would re-architect the legacy of the "promise" on which the arc of philosophic and religious writings depended, but the Trump-effect assumes it is closed and exploitable, short-circuited and self-trolling, void, mocked by the ease of branding ("huge," "great," "winning," "so much better," and so on).

IV

The (im)possibility of this escape, or restoring the legacy of the promise as master-trope, is not the descent into hyper-cynicism of a sort, total decoupling, as Trump does with "climate change," or why I might modify Stiegler's deployment of arche-cinematics as a reading technic that is negentropic by absorbing this cynicism in a hyper-Diogenes inflected manner, which arrives together with an irreversibility Stiegler's discourse need contest as long as the question can be posed – quite correctly, since all the descriptive, analytic, metaphoric, appropriative, corporate or futurist or apocalyptic uses the "anthropocene" has been put to (the geological being the least relevant, a MacGuffin) fail to address the question as performative. That is, as an intervention of sorts in the mnemo-mediatric reformatting of mass cognition and reference. Of course, the logics of escape have all along guaranteed the "Anthropocene" would be, as the conferred name implies, autocidal – for as long as a hyper-elite calculated they could outrun the others, and escape climate chaos, the latter would be accelerated to make that transition, particularly if it coincided with a financially engineered "species split" and techno-eugenicist leap – what a phantom singularity would also name – that coincides, in turn, with a selective mutation.

The heralded populism of *MAGA (Make "Anthropos" Great Again* – as if this were other than a deep fake), which also claims to scoop in the virtual Alt Left (unions, libertarians), is, of course, nothing of the sort, and the holy war against "globalists" – depicted as one-world-order types depriving all of individual sovereignty or control, putatively – is only a sleight of hand, a battle between oligarchic blocks. The obscene benefits of the globalist liberal order funneled to the top ".0001%" creates a billionaire war, with the pirates taking over. Yet who expected the era of climate chaos to yield, eventually, other than climate autocracies in existential competition? The neo-feudal, high-tech rule of autocratic plutocrats globally – each power center left to define and compete or fight out its resources at the poker table. This now perpetual digitized war is not one of choice. Moreover, the fabled overthrow of "globalists" – meaning a rules based international order pretending to respond jointly to emergencies and risks after any Pax America vaporized – gives way to the hyper-internationist cartel of Tech Bros, Real Estate insiders and local ethno-nationalists have no investment, precisely, in the destinies of local populations or their infrastructure woes when they understand the long term "future" both won't exist and will be only for them: of course "we" need accelerate and get this over with since the production of new hybrid homo iterations and A.G.I. agents is already irreversible and the artefaction of higher hominid iterations already claimed and financially engineered.

In its way, the Trumpocene accelerates and discloses the pretenses of the liberal globalist "West" because, in an epoch of climate chaos, of survivalist logics, the logic of escape is the same. It is not accidental that there is endless propaganda about Mars, and getting a backup plan, much as Silicon Valley types are buying helipads in New Zealand (and look for Arctic condos to be the rage by mid-century). That is, the entire subnarrative of the Trumpocene as regards the occlusion of climate change and the vortices of irreversible ecocide is based on the imaginary of flight and escape by the ".0001%" – only now the territorial, gangland style of competing zones and clusters – while the hyper-elite, now oligarchs, double down. The timer is running, but the promised acceleration of AI, gene-editing, biotech hybridities, points not just to a "breakaway civilization" (Catherine Austin Fitts) but a species split, after which the next version will, implicitly, not look fondly at the Neanderthal-like *anthropoi* threatening their biosphere. It may be, from this narrative position to come, the "Anthropocene" will be a mocking name for the current hinge period – when alternative routes had offered themselves theoretically – that of deferred planetary care over-ridden by reactionary denialism that fronts for an escape plan.

Of course, there may be bumps in the road – as when the illusory trope of a single "humanity" shreds into a multiplicity of grades of enhancement, a virtual zoo of such spawned in and fanning out from the coming post A.I. gens ueber caste – hierarchically laddered for inclusion or utility.

The Trumpocene inverts and banishes the "Anthropocene" imaginary while trolling it shamelessly: Trump as wanton *Anthropos*. Yet Trumpism mimes being an evil twin of Stiegler's "escape" from the "Anthropocene" (which, in a sense, everyone wants). It roughly corresponds to the moment, say 2016, when: 1) tipping points are reckoned publicly as passed, conceded in the three-card monte of the "Paris" accords; and 2) irreversibility is met with complete occlusion of the "climate" problematic. (I defer the entire metaphysics of "tipping points" or if they exist.) Thus, I would close out the metaphor of "escape," even as used properly by Stiegler to launch a neganthropic turn against the totalization of the "Anthropocene" and what infects and defines it, against any continuous future project. It is also why I prefer to co-opt the term not as a narrative or dialectic, not as a war in which the "proletarianization" of the senses could be sabotaged or reversed, but as the permanent underbelly of the cognitive and temporal bubble that, today, encircle the fossil of the Anthropocene. It is not incidental that Elon Musk, pushing the inevitable case for Mars as a backup or insurance plan, jests: "Fuck earth!"[80] Fuck all the fetishization of Earth when it is so clearly already fucked – go AI-hybrid, get off this sucker for the species, or some variant, to persist. In what could be called a "first wave," the assumption was of a shared rationality compelled to address auto-extincting behavior. A "save the planet" or sustainability rhetoric would result, as the Paris accords attest publicly. But if the "Anthropocene" invites us to think from geological time, the model we see is rather that of organic life being subsumed, ecological networks eaten through, as if one were dealing with wasp larvae planted in an unsuspecting host that is eaten through at emergence. There never was a logic of preservation, conservancy, restoration, correction at work. The "Anthropocene," accordingly, may come to represent a time of peak hominin population, but primarily those left over from the now cancelled Holocene.

Trump is the first or defining climate president – only that by inversion, by occlusion, by deeming it a point of leverage. After irreversibility is triggered – as it is – the premise is not saving some planet "for all," or all humans, but managing the extinction logics, managing the timing for the next iteration of *homo*. To make the Trumpocene readable one begins noting, of course, that they all "know." *They know* – yet disseminate robo-denialisms strategically. The "Anthropocene" is canceled; climate collapse non-existent – the last great era of mega-extractivism, extending through orbital junk clouds to lunar outposts, weaponized, asteroids, and the weird MacGuffin of "Mars," pinged. All by rhetorical inversion. The takeaway, visible for a decade or more in outline, is broadcast in the brief destructive collusion of Trump and Musk – who had branded as an early climate hero, traumatized by the implications of collapse and devising a Plan B "escape" hatch for the hominid brand, colonizing Mars, mutating, spreading the meme.

When Melville's lawyer narrator ends his exasperated *tale* of the withdrawn Scrivener with the exclamations: "Ah, Bartelby! Ah, humanity!" he plants a backdoor entry, so to speak, to the interpretive traps set. The text apprehends not the wistful mourning of the narratorial puzzlement before Bartleby's withering disappearance, but rather that "humanity," conjured by an apostrophic hailing, had been recategorized and redefined by Bartleby's preference "not" and his definition as scrivener, as copyist without originals, as early veteran of the Dead Letter Office (a training post).

Works Cited

Anderson, Darran. *Imaginary Cities: A Tour of Dream Cities, Nightmare Cities, and Everywhere in Between*. U of Chicago P, 2017.

Anderson, Ross. "Exodus," *Aeon*. https://aeon.co/essays/elon-musk-puts-his-case-for-a-multi-planet-civilisation.

Agamben, Georgio. *Homo Sacer*. Stanford UP, 1998.

Azéma, Marc. *La préhistoire du cinéma: origines paléolithiques de la narration graphique et du cinématographe*. Errance-Passé Simple, 2011.

Badiou Alain. *Manifesto for Philosophy*. State U of New York P, 1999.

Benjamin, Walter. *Illuminations*. Translated by Harry Zohn, Harcourt Brace, 1968.

Bubandt, Nils, and Rane Wilerslev. "The Dark Side of Empathy: Mimesis, Deception, and the Magic of Alterity." *Comparative Studies in Society and History*, vol. 57, no. 1, 2015, pp. 5-34.

Campion, Chris. "Turning Real Terror into Gripping Cinema." *The Telegraph*, 26 Apr. 2004. http://www.telegraph.co.uk/culture/film/3615945/Turning-real-terror-into-gripping-cinema.html.

Chakrabarty, Dipesh. "The Climate of History: Four Theses." *Critical Inquiry*, vol. 35, no. 2, 2009, pp. 197-222.

Clark, Timothy. "Derangements of Scale." *Telemorphosis: Theory in an Era of Climate Change*, edited by Tom Cohen, vol. 1, Open Humanities Press, 2012, pp. 148-67, http://openhumanitiespress.org/telemorphosis.html.

---. "An 'Inhumanist'School?." *Oxford Literary Review*, vol. 45, no. 1, 2023, pp. 142-56.

---. "Toward a Deconstructive Environmental Criticism." *Oxford Literary Review*, vol. 30, no. 1, 2008, pp. 44-68.

Cohen, Tom. editor. *Telemorphosis: Theory in the Era of Climate Change.* Vol. 1, Open Humanities Press, 2012. http://openhumanitiespress.org/telemorphosis.html.

Colebrook, Claire. *Death of the PostHuman: Essays on Extinction.* Vol. 1, Open Humanities Press, 2014, http://openhumanitiespress.org/essays-on-extinction-vol1.html.

The DCDC Global Strategic Trends Programme 2007-2036. UK Ministry of Defence, 2007.

Dean, Aria. "Notes on Blacceleration," *E-Flux Journal*, no. 87, Dec. 2017, https://www.e-flux.com/journal/87/169402/notes-on-blacceleration/.

De Man, Paul. *Aesthetic Ideology.* U of Minnesota P, 1996.

---. "Anthropomorphism and Trope in the Lyric." *The Rhetoric of Romanticism,* by Paul de Man, Columbia UP, 1984, pp. 239-262.

Derrida, Jacques. "Biodegradables: Seven Diary Fragments." *Critical Inquiry,* vol. 15, no. 4, Summer 1989, pp. 812-73.

---. "Cinema and Its Phantoms." Translated by Helen Regueiro Elam (unpublished; original: "La cinéma et ses fantômes," *Cahiers du cinema,* vol. 556, 2001, pp. 75-85).

---. "The Typewriter Ribbon: Limited Ink (2) ("within such limits")." *Material Events: Paul de Man and the Afterlife of Theory,* edited by Tom Cohen et al., U of Minnesota P, 2000, pp. 277-360.

Dimock, Wai Chee. "Literature for the Planet," *PMLA,* vol. 116, no. 1, Jan. 2001, pp. 173-188.

"Drake Well." *Wikipedia,* https://en.wikipedia.org/wiki/Drake_Well.

Dunlap, David. "Putting a Face on the Gulf Oil Leak." *The New York Times,* 4 Jun. 2010, https://archive.nytimes.com/lens.blogs.nytimes.com/2010/06/04/assigment-35/.

Edelman, Lee. *No Future: Queer Theory and the Death Drive.* Duke UP, 2004.

Faulkner, William. *Go Down, Moses.* Vintage, 1970.

---. *Sanctuary.* Knopf, 1978.

Fritsch, M., et al., editors. *Eco-Deconstruction: Derrida and Environmental Philosophy.* Fordham UP, 2018.

Giroux, Henri. "Zombie Politics and Other Late Modern Monstrosities in the Age of Disposability." *Truthdig*, 2 Sept. 2012, http://archive.truthout.org/111709Giroux.

Ghosh, Amitav. The Great Derangement: Climate Change and the Unthinkable (University of Chicago Press, 2016.

Glissant, Eduoard. *Faulkner, Mississippi*. U of Chicago P, 1996.

Hägglund, Martin. *Radical Atheism: Derrida and the Time of Life*. Stanford UP, 2008.

Hamilton, Clive. "Utopias in the Anthropocene." *Plenary Session of the American Sociological Association*, Denver, 17 Aug. 2012, http://mahb.stanford.edu/wp-content/uploads/2012/08/2012-Clive-Hamilton-Denver-ASA-Talk.pdf.

Hansen, James. *Storms of My Grandchildren*. Bloomsbury, 2009.

Harari, Yuval Noah. *Nexus: A Brief History of Information Networks from the Stone Age to AI*. Random House, 2024.

Hedges, Chris. "The Pathology of the Rich: Chris Hedges on Reality Asserts Itself." *The Real News*, 5 Dec. 2013, http://zeropointportal.us/pathology-rich-chris-hedges-reality-asserts-pt1/.

Hoens, Dominek, et al., editors. *The Catastrophic Imperative: Subjectivity, Time and Memory in Contemporary Thought*. Palgrave Macmillan, 2009.

Kaplan, Robert. *The Nothing That Is: A Natural History of Zero*. Oxford UP, 1999.

Latour, Bruno. "An Attempt at a 'Compositionist Manifesto,'" *New Literary History*, vol. 41, 2010, pp. 471-90.

Lynes, Phillippe. *Futures of Life Death on Earth: Derrida's General Ecology*. Rowman & Littlefield, 2018.

Malabou, Catherine. "Interview: On the Clitoris, AI, Anarchy, Marx…." *Crisis and Critique*, 6 Mar. 2024, https://podcasts.apple.com/us/podcast/catherine-malabou-on-the-clitoris-ai-anarchism-hegel/id1617435046?i=1000648277260.

Marché, Stephen, "An Epidemic of Facelessness." *The New York Times*, 14 Feb. 2015, https://www.nytimes.com/2015/02/15/opinion/sunday/the-epidemic-of-facelessness.html.

Miller, J. Hillis. "Literary Study in the Transnational University: The Fractal Mosaic." *Black Holes*, edited by J. Hillis Miller and Manuel Assensi, Stanford UP, 1999.

---. "Ideology and Topography: Faulkner." *Topographies*, edited by J. Hillis Miller, Stanford UP, 1995, pp. 192-215.

Morrison, Toni. *Playing in the Dark*. Harvard UP, 1992.

---. *Sula*. Random House, 2006.

Negarestani, Reza. *Cyclonopedia: Complicity with Anonymous Materials*. Repress, 2008.

Naremore, James. "Love and Death in A.I. Artificial Intelligence," *Michigan Quarterly Review*, vol. 44, no. 2, Spring 2005, pp. 256-84.

Novak, Phillip. "'Circles and Circles of Sorrow': In the Wake of Morrison's *Sula*," *PMLA*, vol. 114, no. 1, Jan. 1999, pp. 184-193.

Paczkowski, John. "Code/Red: The Best Elon Quote, Ever." https://www.vox.com/2014/10/1/11631474/codered-the-best-elon-musk-quote-ever.

Praschl, Peter. "The Global Village Has Become a Nightmare." *WorldCrunch*, 5 Oct. 2014, http://www.worldcrunch.com/opinion-analysis/the-global-village-has-become-a-nightmare/global-village-failure-world/c7s17120/#.VDJIokuxDwI.

Purdy, Jedediah. "Anthropocene Fever," *Aeon*. 2015. https://aeon.co/essays/should-we-be-suspicious-of-the-anthropocene-idea.

Read, Justin. "Unicity." *Telemorphosis: Theory in the Era of Climate Change*, edited by Tom Cohen, vol. 1, Open Humanities Press, 2012, pp. 121-48, http://openhumanitiespress.org/telemorphosis.html.

Rogers, Adam. "The World Has Never Seen Anything Like What's Happening at the Equator Right Now." *Mother Jones*, 17 Sept. 2018, https://www.motherjones.com/environment/2018/09/the-world-has-never-seen-anything-like-whats-happening-at-the-equator-right-now/.

Ronell, Avital. *Crack Wars: Literature Addiction Mania*. U of Nebraska P, 1992.

---. "The Rhetoric of Testing." *Stupidity*, by Avital Ronell, U of Illinois P, 2002.

---. *The Test Drive*. U of Illinois P, 2005.

Rotman, Brian. "The Alphabetic Body," *Parallax*, vol. 8, no. 22, Jan.-Mar. 2002, pp. 92-104.

Scranton, Roy. "Learning How to Die in the Anthropocene." *The New York Times*, 11 Nov. 2013, http://opinionator.blogs.nytimes.com/2013/11/10/l earning-how-to-die-in-the-anthropocene/?_r=0.

Stephenson, Wen. "From Occupy to Climate Justice." *The Nation*, 24 Feb. 2014, http://www.thenation.com/article/178242/occupy-climate-justice.

Stiegler, Bernard. "Nanomutations, Hypomnemata and Grammatisation." Translated by Georges Collins, *Ars industrialis: association internationale pour une politique industrielle des technologies de l'esprit*, 2006, http://arsindustrialis.org/node/2937.

---. "The Organology of Dreams and Arche-Cinema," *Screening the Past*, June 2013, http://www.screeningthepast.com/2013/06/the-organology-of-dreams-and-arche-cinema/.

---. *Technics and Time, 3: Cinematic Time and the Question of Malaise*. Translated by Stephen Barker, Stanford UP, 2010.

---. *Uncontrollable Societies of Disaffected Individuals: Disbelief and Discredit*. Vol. 2, Polity Press, 2013.

Chakravorty, Spivak Gayatri. *Death of a Discipline*. Columbia UP, 2003.

Sundquist, Eric. *Faulkner: The House Divided*. Johns Hopkins UP, 1983.

Taibbi, Matt. "Trump the Destroyer." *Rolling Stone*, 22 Mar. 2017, https://www.rollingstone.com/politics/politics-features/trump-the-destroyer-127808/.

---. "Why Aren't We Talking More about Trump's Nihilism?." *Rolling Stone*, 1 Oct. 2018, https://www.rollingstone.com/politics/politics-news/trump-white-house-climate-change-731440/.

Timofeeva, Oxana. "Ultra-Black: Towards a Materialist Theory of Oil." *E-Flux Journal*, no. 84, 2017, https://www.e-flux.com/journal/84/149335/ultra-black-towards-a-materialist-theory-of-oil/.

Townsend, Mark, and Paul Harris. "Now the Pentagon Tells Bush: Climate Change Will Destroy Us." *The Guardian*, 21 Feb. 2004, http://www.theguardian.com/environment/2004/feb/22/usnews.theobserver.

Vitale, Francesco. *Bio-Deconstruction: Derrida and the Life Sciences.* State U of New York P, 2018.

Wagner-Martin, Linda, editor. *New Essays on Go Down, Moses.* Cambridge UP, 1996.

Wittenberg, Judith Bryant. "*Go Down, Moses* and the Discourse of Environmentalism." *New Essays on Go Down, Moses,* edited by Linda Wagner-Martin, Cambridge UP, 1996, pp. 49-72.

Wood, David. "On Being Haunted by the Future." *Research in Phenomenology,* vol. 36, no. 1, 2006, pp. 274-98.

---. "Specters of Derrida: On the Way To Econstruction." *Ecospirit: Religions and Philosophies for the Earth,* edited by Laurel Kearns and Catherine Keller, Fordham UP, 2007, pp. 264-88.

Žižek, Slavoj. "ISIS Is a Disgrace to True Fundamentalism." 2013, https://zizek.uk/2014/09/03/isis-is-a-disgrace-to-true-fundamentalism/.

---. "Nature and Its Discontents." *SubStance,* vol.117, no. 37.2, 2008, pp. 37-72.

---. *The Puppet and the Dwarf: The Perverse Core of Christianity.* MIT Press, 2003.

Notes

1. See Clark "An 'Inhumanist' School" 142-56. Clark – who more than anyone perhaps pressed "deconstruction" into belated engagements with the climate-extinction algorithms (and authored a still classic essay on "Scale") – questions the present author and Claire Colebrook's joint publications as an "Inhumanist" school at the margins of an eco-deconstructive project. It (this "school") would be one that pivots to de Man from Derrida to activate tools weirdly suited to the evolving non-present of what we might call, for now, the "Trumpocene." Clark's shrewd and generous review ("an unignorable purgative for workers in the environmental humanities") registers that this "school" toggles between archae-graphics and digital AI – wherein the "*in*human" aspect of language (as Benjamin terms it), "material" marking and aural inscriptions, displays itself openly as the digital mnemotechnic streams and unfolding AGI entities usurping sentience and embodiment: "The most salient and important part of the intervention made here is surely this: that many responses to threats of civilizational and environmental collapse entail modes of thinking complicit with the destructiveness in the first place" (143). Clark pulls back, as one legitimately might, from the recurrent premise that "tipping points" are now programmed in – moreover, guaranteed by the default hermeneutic reflexes. Clark critiques "their slightly excessive eagerness to claim that the truly decisive climatic 'tipping points' have long been passed… and 'the irreversibility of ecocide [is] the invitation to a terminological reset.'" Yet perhaps the election of Trump will have clarified the premise, in which Clark maintains wriggle room (the "truly decisive" events seem to hedge from less "truly decisive" ones). The dismissal of "climate change" publicly as hoax (again) and "drill, baby, drill" guarantee accelerating cascade events and mirror a return to oil and pratfall international conferences (that the Saudis have decided to repeatedly blow up), makes some things clearer. Clark has assiduously probed the resistance of "deconstruction" – along with its transparent readiness or connection to – the horizons of climate chaos, which Derrida had passed on (or, I suggest, occluded).

2. See Timofeeva.

3   See Dean.

4   Yuval Noah Harari, famous for *Sapiens* and *Homo Deus*, seeks to spawn a pan-epochal perspective on the "human" trajectory. *Nexus: A Brief History of Information Networks from the Stone Age to AI* (2024) supplements this series with the shift of focus to what he terms "information" technology but might be better termed *mnemotechnic* regimes. The focus for this new scan will be named "information," and then *networks*, but it is all about the agency of *inscription* technologies, memory technologies, their institutionalization and appropriation, and the "stories" or fictions that are inextricable and shape *realities* decisively – all of which become vulnerable in a "time" of climate disinhabitation, coming resource wars (now including micro-chip production, the moon, asteroids, neighbors) and the Cambrian explosion of AI variants. Harari does not focus on the specific agency of these *hyper-material* regimes, currently digitally transposed and fused. One might locate this volume's sketches within a current *hinge* episode in *this* trajectory, an imperiled transformation, which has deposited us in a rising post-literate and pre-literate mnemo-social escarpment: what occurs to "reading" in the transition from an era of the book stripped of certain tribal contracts, programmatic loops, hermeneutic reflexes, identificatory shufflings – one which grasps that "era" and those preceding as an evolving ensemble of mnemotechnologies, now passed into the digital and early AGI ubiquity? Is complaining of a transition to algorithmic dominance a less productive strategy than examining the *hyper-materiality* itself of modes of "information" shaped by and in these technologies – and the technicity that threads them? This blind spot seems untransgressable and lethally transgressive.

5   "Drake Well," https://en.wikipedia.org/wiki/Drake_Well.

6   The US in its political and digital theatrics presents a harsh caricature of these threads. Clown-genius Musk, and video-game bot, who used to cry at the thought of human extinction and present as savior of climate ruin (and, if not, Mars escape), now prances on X as chief chaos agent, allied with those who deny "climate change" (MAGA Republicans, Apocalyptic "Christians") and any problem with oil until it runs out. All of this programmed by the saturnine Peter Thiel whose initial embrace of Republican neo-fascists fantasized a libertarian outcome, surfing the crypto-Christian coup of the Supreme Court, and who resonates to Marc Andreeson's stoner manifesto, "The Techno-Optimist," welcoming the retirement of humans as the oceanic mutations of AGI arrive. The not so discreet subtext for this *triad* – white Evangelicals, "Trump" and the Mega-Tech Bros – is not that climate chaos is not real but that it is, *too late* to bother trying to turn around. The extractivist looting, the engineered species splits and the assertion of a supreme caste to come

(Musk's ruling, "Alpha"-male cohort) proceed apace. This, so that the hybrid-engineered *anthropoi* to come, gifted longevities, replaceable organs, networked implants and quantum interfaces, outpace the more general write-off or triaging of countries, area, and peoples incrementally.

7   One has moved far from the Nietzschean question that disbands "the human" as an epistemographic primitive to be hyperbolically recast to survive. Instead, literalization reigns, so that a "singularity" uploading individual psyches to evade organic death, or the expectation of an AI godlet to come or hybrid human caste appropriates the very different aims of the first. Instead of a hyper-mensch, the "post-human" intends to outwit mortality, supplement or replace the body, upload select "minds," make the present of *last-man* culture frozen and indefinite against time-space, the organic, corruption.

8   What draws me to this trope of a *malignant narcissism* is that such "malignancy" can represent a counter-*pharmakos* that flows beneath received tropological spells, drips from and through our more communally conferred and regulated narcissisms, the composition and defense of "we's," whose current normativity is unwittingly malignant and produces mass extinction events. Ronell goes places, such as to a black ooze melting through gender wars into a non-site comparable to the deliquescence of ink itself, a crypto-materiality which ruptures the more sunlit narcissisms that operate, today, as maintenance systems for doomed knowledge formations and twentieth-century idioms. Since Derrida not only veered away from the unfolding twenty-first-century biomaterial mutations and aporia becoming manifest before his death (the wayward term "Anthropocene" being introduced around 2003), it is more interesting to ask, today, if her writing apparatus had drifted, *avant la lettre*, precisely into this zone.

9   See Malabou.

10 This sketch was originally presented in Rio de Janeiro, hence the tie-in to a passing reference to the set of Hitchcock's *Notorious*, in which "Rio" is marked as a circular or cinematic pun, a telepolis or the screen itself.

11 See Ronell, *Crack Wars*.

12 Indeed, *Stupidity* opens with it five times in its first three sentences, two in its first: "The temptation is to wage war on stupidity as if it were a vanquishable object – as if we still knew how to wage war or circumscribe an object in a manner that would be productive of meaning or give rise to futurity" (3).

13 See Faulkner, *Sanctuary*.

14 See Faulkner, *Go Down, Moses*.

15 We will ignore the slippage into Hitchcock, on whose *Rope* Ronell has interesting comments in *The Test Drive*, in a writing in which telephonics and cinematic teletechnics converge about a certain wheel, circuit or dial tone; in which scissors lodge in the back of the intruder; in which ocularcentrism is, from the start, suspended as the graphematic scandal of what is too obvious – that the effect of living is animation, that the black ooze of Emma is instantaneous, exceeds its Flaubertian discharge, lies in the "visible" order of the prefigural mark.

16 In "Go Down, Moses," the final and titular tale in *Go Down, Moses*, we hear of the lawyer Gavin Stevens that the law was his "hobby" and his "serious vocation was a twenty-two year old unfinished translation of the Old Testament *back* into classic Greek" – that is, a *translation* of the biblical text's (English) translation as though "back" not into an originary language but into the aesthetic language which, nonetheless, precedes Mosaic authority.

17 It will take a century (we are not there yet by this count) for "Faulkner's" gesture to be apprehended: that everything here pretended to in the ritual of the "literary" is closed, never was as pretended to by the parlor-room class of Horaces, by the "law," henpecked.

18 See Praschl.

19 Žižek kind of reads ISIS with the same disappointment as he does liberal Marxists – as not "radical" enough – "ISIS Is a Disgrace to True Fundamentalism" (2013). He cites Nietzsche's "last men": "Long ago Friedrich Nietzsche perceived how Western civilization was moving in the direction of the Last Man, an apathetic creature with no great passion or commitment. Unable to dream, tired of life, he takes no risks, seeking only comfort and security: 'A little poison now and then: that makes for pleasant dreams. And much poison at the end, for a pleasant death. They have their little pleasures for the day, and their little pleasures for the night, but they have a regard for health. 'We have discovered happiness,' say the Last Men, and they blink.'"

20 Osama bin Laden saw this, toward the end, when he shifted the rhetoric of Al Qaeda as one mobilized against hyper-industrial modernity's generating climate change and mass culling to come.

21 Paul de Man, "Anthropomorphism and Trope in the Lyric," in *The Rhetoric of Romanticism* (Columbia UP 2000), 241. Since "anthropomorphism" is not a trope, it does not give or project face. It also cannot enter a substitutive chain of possible meanings that are in transformative play. Rather, de Man says in "Anthropomorphism and the Lyric" that it in effect *freezes* that chain, totalizes and disables it at

once – pretends to live in the spectrum of figural multiplicities yet, all along, dissociates from what are, by virtue of it, drained of aura. The proper name *Anthropos* is without any *definitional* in advance of itself to cite or differ from. As Xenophanes notes, if for man the gods have human form, the same goes for the gods of horses or goats.

22 See Marché.

23 As such, the birds cannot be *apocalyptic*. Their invasion as a warping of temporal logic implies a folding in of the frame, without outside. These birds are not animals but *technemes*, allied to machines, telegraph wires, pecks, the prehistorial and post-anthropomorphic, the cinematic as such. *An Earth*, after all, is what the camera will always (also) be gazing at, be recording – even when a simulacrum Earth, as with sets of anthropomorphic monuments made of stone and steel. But *nature* was always other than maternal, an anthropomorphized "she," the shot's representational claim. "She" is something else, *proactively* mimetic, a mimesis without model or copy, much as species alter ceaselessly according to the technicity of an environs or for camouflage or shape-shifting, adapting proleptically, as an animal or coral sea creature or insect assumes camouflage before a predatory other when it cannot "see" itself to be like the mimicked twig or rock or leaf. This cin-animation of "Earth" is the paradoxical counter-world to the passivity of "globalization."

24 See Lee Edelman's powerful reading of *The Birds* as an abruption of futurity and reproduction in *No Future* (2004).

25 The contradiction between a tempophagic present and a future population whose reserves have been taken and who are saddled with megadebt before extreme climate change has been used and characterized variably without any indication a collective ethics is triggered – what locates the idea of future generations, to begin with, in a hostile, competitive or lethal gaze as well. Thus, James Hansen uses the figure to mobilize his account of doomed carbon imbalance and the failure of politics in *Storms of My Grandchildren* (2009). Yet we already have proleptic *defenses* of the present generations' crimes in this regard, firing back in anticipation that they, the latter, would have done no different and probably still won't be doing so.

26 Indeed, references to *oil* are systematic in Hitchcock, this *black liquid* source of prosthetic energy and hence the electric light and all vehicles of transport – the fossil residue of the birds' prehistorial forebears, dinosaurs, cycling back like Norman's bog.

27 Mel Gibson's *Signs* inverts *The Birds* precisely – as attacking space aliens beaten off by family and faith, by the human. The exact opposite of Hitchcock – these birds are domestic, of the frame or background, invisibly visible, attacking the eye, driving from the house

of metaphor, assaulting the biopolitics of the image or how it has been constructed.

28 Among the avatars of these birds in Hitchcock is not only a menagerie of *animemes*, especially cats, but black suns and marbles, airplanes. Or the fly on "mother's" hand – addressed by her through Norman in his *cell*, when "mother" would show, by being still before the viewer, how she wouldn't hurt a fly. That *fly*, exoskeletal, cannot be hurt by "mother" in any event – since it is a flying hole or tear in representational orders, like the black sun shot at in the first *Man Who Knew Too Much*, which inverts the perceptual premises of "image" in its entirety, in its "general semantics," and opens other temporalities and wormholes.

29 See Miller, "The Fractal Mosaic."

30 See Miller, "Ideology and Topography."

31 For a critical interrogation of "materiality without matter," see Derrida, "The Typewriter Ribbon."

32 See Faulkner, *Go Down, Moses*.

33 That "Faulkner" is entirely alert to this auto-inscription and archival gamble (or erasure) is marked not only by his covert citing of titles from his own production in the performative-theoretical evisceration of the history of "the book" in section four of "The Bear" (notably *Sanctuary* and *The Sound and the Fury*, referenced to a sort of anti-God "He" on page 272), but the use of the term *"catafalque"* – a double fall, a going down of *"fa(u)lque(ner),"* a bier on display across the disarticulations of *Go Down, Moses* – in the titular final short tale, "Go Down, Moses": "in formal component complement to the Negro murderer's catafalque: the slain wolf" (364).

34 Eric Sundquist notes of *Go Down, Moses* that "Faulkner is moving further away from his own early 'modernism' and placing himself, deliberately or not, more clearly in the tradition of classic nineteenth-century American fiction" (133), that is, a regression to a type of naturalism, or mimeticism, that is not without a sentimental falling off as well. He adds: "Go Down, Moses may be Faulkner's most honest and personally revealing novel, even though it is clearly not his best" (192). That is, the more "honest" it is, the more mimetic it is, yet as a falling off into sentiment merely, the weaker, or hence more dishonest it also is, and so on. In *Playing in the Dark*, Toni Morrison complains that, typically, as soon as Faulkner is seen as treating "blacks" primarily, the writing is itself cast off as the author's decline. We will suggest, below, a still more complicated relation between the problematic of blackness, animals, falling or going under and Mosaic "authorship" than either view represents.

35 An extended scrutiny of "Sophonsiba" would link her not only to the eviscerated, dreaded and wholly empowered domain of "woman" in the book, but with sound, rhetoric, aesthetics, cognition (in the mock-allegorical allusions of the name), in short, the Greek.

36 The variously deployed figure of "trace-chains" – at once mnemonically inflected and bearing the terms of (t)race and enslavement – will recur across the different texts, beginning with Sophonsiba's association with sound or the ear in "Was": "the earrings and beads clashing and jingling like little *trace* chains" (10, my italics). The term will crisscross mnemonic constellations and scriptive nodes as what stand to be modified or altered in the production of transformed historial, proto- and post-Mosaic "events." Compare to the recurrence of "trace-chains" in "The Bear" (245).

37 One of the sites in which the problematic of the "aesthetic" is played out is around the narrative use of Keats's Grecian urn ode in relation to Uncle Hubert Beauchamp's self-borrowed-against and emptied-in-advance "gift" (to Isaac) of the "tin coffee-pot" (297), the "legacy" in Chapter Four of "The Bear." The text performs the dispossession of an entire legacy of the "aesthetic," of aesthetic ideology tout court, yet by its own agency.

38 While the work equivocates between the façades of Greek and Roman and Hebraic – that is, unreliable markers of the "aesthetic," the churchly-mimetic and the originary law – it does so in an unstable way that discloses, at first, an Egyptian motif beneath each, as if these options did not indicate historical periods but constructed variants of an underlying marking system, almost "unreadable" (to whites) yet as hieroglyphic as the letter "delta" that appears on the page of "Delta Autumn," like a pyramid, or the distortions to the story of Joseph that Mollie Beauchamp introduces in her chant: "Roth Edmonds sold my Benjamin. Sold him to Egypt. Pharaoh got him – – " (353).

39 See Walter Benjamin, "The Task of the Translator," in *Illuminations*, 80.

40 Such a virtual shift corresponds in ways to what de Man called a passage from reading language as (mere) tropes, systems of displacements, to "another conception of language" as a kind of materiality precedent to figuration, which he terms "performative" in *Aesthetic Ideology*, 132. De Man adds a note concerning this "one-way street" of Benjamin: "that process ... is irreversible. That goes in that direction and you cannot get back from the one to the one before" (133). De Man's insistence on a passage, direction, shift, movement or translation occurs like a drumbeat in *Aesthetic Ideology*: "pass from that conception of language to another conception of language" (132); "the passage from trope... to the performative" (132); "there

is a single-directed movement" (133); "there is a road that goes from this notion of *Schein* to the notion of materiality" (152).

41 See Philip Novak, "Circles and Circles of Sorrow": In the Wake of Morrison's *Sula*. PMLA, 114:2 (March 1999), 191.

42 The recent recyling of the motif of genetic determinism and absence of free will by protégés of Richard Dawkins's selfish-gene trope ignores that Dawkins's real contribution is where that logic manifests as memes, as linguistic algos and mnemonic installations; thus, the import of this *literalism* (that it is genetic) serves its own anaesthetizing role amid the current dangers. If genetic determinism loses its narrative role to algorithmic predictivity, it is also the case that Dawkins real contribution and subject would be *memes*.

43 And perhaps it turns out what we named "irony" itself, decoupled from staged intention, is the interpretive containment of the perspective of the "inhuman," which involves en route language and digital semiosis implicitly (so, at least, Benjamin).

44 What de Man perhaps meant by "literariness" would not be our usual associations of the term, say, with style, narrative convention, tropological networks – fictions. I would like to reconnect several moves that appear in de Man by locating in what he calls "aesthetic ideology" a cause in the advance and acceleration of ecocide today. In this regard, what de Man means by "literariness" is a non-site in which interpretive regimes have not been installed, face not instituted (again by accord), various mimetic automatisms not yet triggered. What there is, here, is an organization of inscriptions that has neither life nor presence. This has resonance with where Bernard Stiegler returns us to the figure of arche-cinema – older than Derrida's arche-writing, pre-letter and pre-scriptive – which he both references to the cave paintings' cinematic projection and presents as the apparatus out of which the technics of "consciousness" is produced. Being in a movie or not is not easy to decide, as with dreams or hallucinations, much as "perception" itself, convinced it is of the moment, forgets the technics that co-instituted it.

45 Thus, it is necessary and has been for a while to *index* every rupture and impending or reactive or weaponized trauma – from geopolitical to financial to, of course, technological outbidding on all fronts linked to AGI, to the co-opted vocabularies and self-sabotaging terminologies; everything should be indexed, today, not only to a climate change *unconscious* spawned by all the rationales to evade, occlude or defer a proper response, but to an underlying *panic of reference* itself. It is also fundamentally the rip through entire conceptual and semantic hives predicated on the legacy epistemographies of a Holocene or even just post-war, twentieth-century conceptual anchors. After the fossil fuel cartels, crypto-evangelist,

techno-optimists, and techno-feudal, financial overlords, and your now smug array of neo-fascist populists, capitalizing on its being "too late" to serious correct the acceleration; after these actors, the greatest obstructors of engaging or addressing climate ruin at the appropriate time would be the humanist commandments of "cultural studies" and the political literalisms and categories of "social justice" as a final appeal. To engage this horizon required first breaking a dominant, passive mimetic ideology, now massively co-opted and driven by algo swarms.

46   In Hitchcock's *Spellbound*, again, the Rome to which all roads lead, including currency itself (petro-dollar), is shifted post-war to New York City, the "empire" state and media capital – portayed as a sort of Grand Central Station of cinematic tracks converging, transport, coming and going?

47   See Derrida, "Biodegradables."

48   Or in an older idiom, if text deconstructs *itself*, then the Platonic concepts generated by the writing of the dialogs – in fact, the *eidos* nowhere coheres or is endorsed in the writing – but is perpetually produced by the misreading and inscription evading "present" of hermeneutic regimes and epochal epistemes defined by their blindzones.

49   An entire essay might be needed to explore the critical innovations of the signal collection, titled *Eco-Deconstruction: Derrida and Environmental Philosophy* (2018) – a stand-out volume with a majority of first-rate and diverse contributors, hobbled by a few ticks bound to the academic premise of "environmental humanities" (really?) and, frankly, "eco-deconstruction" where the yearning to habilitate the family names ("deconstruction," were this now identifiable; "Derrida," of course). Not, that is, Derridean writing interrogated from the era of climate chaos and tempophagic entropies – which is not in Derrida's text. One remarks the oddity of such sophisticated critical theology getting readers beyond that of its own circuit, if it wished to intervene in the climate-conceptual-complex – and judging from three Derrida conferences I am aware of in 2024, it did not induce any reset. Two essays stand out, those of Michael Marder and Claire Colebrook, for criticizing Derrida's occlusion of interest in and allergy to the "eco" per se. (In a private exchange with a young editor of the volume at a "Derrida Today" conference at the time of publication, I was told that Marder's essay raised consternation among the programmatic editors, but then Professor Frisch told me a scientist he knew assured him the climate emergency – call it what you will – was not a long-term problem.) It isn't necessary to say that my critique of Derrida and his aftermath, of the legacy clusterfuck of fused and competing heirlets and legatees, recognizes

the full spectrum of inventions, counter-poisons and performative seedings that constitute a philosophical-historial "event" – but equally, that the behavior of mourning and Derrida's own occlusion of the climate-chaos horizon required a very different shattering and rewiring of select interventions to pair with the digital capture of inscriptions and sabotage of attention and literacies that the climate-extinction "era" fused with (all of two decades so far, an infant in every way). The primary self-sabotaging of the "eco-deco" portfolio, aside from what its title betrays, is propping itself up with the genealogical tricks that animate Vitale's "discovering" of unfinished notes from the '70s showing that Derrida would have been there all along, and originarily, and Lynes's reliance on the *Life-Death* figure to rework the adjacency of Derrida to the climate-extinction emergency – the auto-immune suicidism transferred from psycho-socio-political into *biotechnic* orders. In the first, the whole exercise of discovering an edited text that authorizes a looked for meaning, shows a weak necessity for continuity in the management of the proper name, "Derrida," and what would be an almost rote target of Decon 101 practice – removing such genealogical tricks from the action of writing. As for *Life-Death*, the figure lay there for forty-odd years and would have been better framed as *Death-Life* from the start: it is no different than that trace ever was, and as a forensic tool or point of appeal, still seems attached to mourning. The official management of the brand "Deconstruction" should be replaced by scions that hadn't read or been mentored by themselves.

50 Stiegler remains a marker and event in this parade, as the one who – radicalizing Derrida's technics and posting it as pre-scriptive and linked to the unfolding both of terminal climate chaos and the usurpation of will and attention by the digital pre-configuration of A.I. totalization and algorithmic capture – pivoted to plunge straight into the black sun of the Anthropocene extinction manual and addictogenic bleed out, accepting a responsibility to strike into the complex knot. He called this the attempt to be "worthy" of the legacies he would precede to alter this capture, to "escape" the Anthropocene doom trap. This, at the time or decade (tic-toc) when the window to switchup or outwit the vortex would close (and, frankly, had already unremarked).

51 The proposed title updates and rewrites the kitsch blockbuster, *2012*, in which catastrophic climate change destroys all continents in twenty-four hours – to fit it all within a movie: well, in our proposed film, the temporality would be somewhat reversed. Instead of a massive opportunity for the spectacular destruction of all terrestrial surfaces (and almost all humans), condensed to a single day, we find a split within.

52 This phrase marks both Stiegler's reprioritization of "the aesthetic," the latter's non-organic technogenesis of perceptual "consciousness" and its *ill*, contamination, default and prescribed vulnerability to capture – and not just preemptively by the NSA or databots. If the "organs" of the senses are composed differently in different technical epochs, they have always been subject to being systemically hacked. Outsourced memory is not returned to the dependent, with a consequent loss of "knowledge" and specifically knowledge of life and how to live (savoir faire). The "proletarianization of the senses" is associated with a "short-circuiting" which today is allied for Stiegler to the collapse of care and attention, mass "dis-individuation" and the accelerations of mafia cultures and ecocide. In writing on arche-cinema, he elaborates: "it is the primary and secondary identification processes, which constitute the condition of formation of the psychic apparatus, and therefore the condition of production of libidinal energy, that are effectively short-circuited." Yet the phrase is not just a hyper-industrial experience of digital, last-man culture, where it goes hyperbolic. This "short-circuiting" would have arrived with the advent of *tertiary retention*, or when, instantly, it remarks itself – a negative condition of the evolution of technical objects that seeks to mark, and negate, its recent form. In a way, when tertiary retention hits the mirror stage and takes a selfie, short-circuiting is triggered. The term "Anthropocene" operates in this way, which accounts for its surge of popularity – it is a short-circuiting selfie of a sort, a closed circle. If the "proletarianization of the senses" leads into the core of a spell, of which the inability to see what is before one in "climate chaos" is an example and prize (media-streamed denialism), it leads us back to where, in Stiegler, the apparatus of this spell hovers – what he names arche-cinema.

53 This mise en scène, in fact, is pretty much verbatim implied by a Chris Hedges, the American activist and writer. It brings home the dilemma of utopist politics, whose time – like democracy or Enlightenment memes – appears to be closing, and recedes before resource wars, mega-drought, agricultural collapse and the sixth mass extinction event underway. Here is Hedges: "Corporations are, theologically speaking, institutions of death. They commodify everything – the natural world, human beings – that they exploit until exhaustion or collapse. They know no limits. There are no impediments now to corporations. None. And what they want is for us to give up. They want us to become passive. They want us to become tacitly complicit in our own destruction.... I think they know it's going to be toast. And I think they think that they're going to retreat into their, you know, gated compounds and survive it. And they may survive it longer than the rest of us, but in the end, climate change alone is going to get us.... [They], if left unchecked,

will ensure the extinction of the human species. It may already be too late, of course." See Hedges.

54 Perhaps something like *accelerated* extinction must now be affirmed, with a caveat. That this transition is viewed *positively*, and leveraged to retire only the current iteration of hominins. Alternately, this logic resets and re-reads *everything*. Bruno Latour deems the Modernist parenthesis at fault, citing Benjamin's confused angel, who, in constantly looking back to redeem the catastrophe of the past, cannot look forward to the one that he, himself, is.

55 This is why *Avatar's* romanticization of the Native Americans styled "Na'xi," rooted in a natural world that itself lives organically (like Lovelock's *Gaia* trope), is less regressive than cynical, since it is ultimately a critique of the kind of organicism that feeds the common imaginary's appetite for romantic organic metaphors of return to systemic wholes. It is the fantasy of the military – a critique which the notion of the *avatar*, the legless film viewer inhabiting the athletic body of the hero soldier's projection into the Na'xi world, leaves open.

56 Latour suggests we give up words like "future." He tropes this as the Moderns' inability to turn from his catastrophic past, always wanting to redeem it (like Benjamin's stupid angel), rather than turning to see where this leads. One has the distinct sense that our theorists, if faced with the choice between keeping their twentieth-century conceptual investments and costing future generations options for existence or giving up the former to give the latter a chance, would choose the first every time. This is the broad impasse of utopist politics today, which critics like Clive Hamilton and Dipesh Chakrabarty, among others, have posed: as a social model in which human struggles with one another are the encompassing horizon, it has in fact contributed to the acceleration – rather than, as it would wish, being free of contamination of an imperial machine like "capitalism" (which tends to make us victims and free of responsibility). And it had waited too long to for its sell-by date before the import of cascade events and the crafting of algorithmic populations that nullify agency appear irreversibly attendant on an orchestrated species split with an eye, precisely, on weaponized climate chaos.

57 When Ian Baucom attempts to recuperate Dipesh Chakrabarty's intervention, to save or restore "post-colonial" historicism in the era of climate change, and does this by assuming the "third history" of geological and biomorphic mutation can be added on, assimilated, to the utopist agenda in a sort of continuity, he misses the point, and re-enacts the hermeneutic refold or restoration that, like ecologism itself, fuels and advances the acceleration (such are the times we're in).

58 In decades (so goes the logic), the survivor caste will have evolved through monopolizing nanotechnologies and resources, and will look back on our present as messy Humanity 1.0 – *Neanthropos* – when all those mixed genetic pools and diseases and hangovers from arcane ideologies were rampant and dangerous. Look, they will say, they almost destroyed the earth, and we'll never get out of that fully – they were like locusts, while we saved the species and allowed it to advance. A certain faux Darwin would be cited for approval.

59 The argument is that Derrida occluded "climate change" from his late writings as, it turns out, he occluded "cinema" from his work almost entirely – no-go zones, zones that interfered with the rhetorical premise that a "deconstruction" could occur.

60 See Rogers.

61 See Naremore.

62 See Derrida, "Cinema and Its Phantoms."

63 See Anderson 5.

64 These notes are fragments of a longer monograph, a gallery of readings of specific shots or images that complicate the agency of cinema (and photography) in the arc of extinction that the term "Anthropocene" implies. My warm thanks to Alexander Strecker whose critical comments helped shape an early form of this essay.

65 As Matt Taibbi writes of the astonishing nihilism of Trump's death-drive: "A policy that not only recognizes but embraces inevitable global catastrophe is the ultimate expression of Trump's somehow under-reported nihilism.... Obese and rotting, close enough to the physical end himself (and long ago spiritually dead), Trump essentially told his frustrated, pessimistic crowds that America was doomed anyway, so we might as well stop worrying and floor it to the end...." See Taibbi, "Trump's Nihilism."

66 Where cinema in this *phase* partakes of AI and trades in ghost representations (the screen actor per definition is a "skinjob" or "Replicant" in *Blade Runner's* sense), the least implication of identification with an unliving and undying simulacrum who is "more human than human" and more real to us than bodily existents (we apparently bond with TV faces or characters as "friends" more than actual humans).

67 See Campion.

68 In Hitchcock's marking system "Piccadilly *Circus*" tags the "Pi" with his cipher for cinematics (3.14, obverting the ubiquitous 1and 3 (thirteen), while "circus" amplifies the wheels and filmloop of a circular dilemma and system – what the "time-bomb" (apparently

always too late or too early) on the *bus* (like a train, trope of machinal *transport* in Hitchcock) would alter, transform, down to the biological entities in the aquarium tank as "screen."

69 What emerges is a film that does not coincide with the sociologist's commentary on ethical impasses and that, in the end, he sees as without hope since they are premised on "efficiency" – that is, one is within a self interpreting system like the revolving jails, a film-loop. It is, after all, just such "biopolitical" theses that are given to the socially conscious and liberal students in Padilha's next film, a feature film the opposite of *Bus 174*. In *Tropa de elite* (2008), we are even given the "biopolitical" model in association with the name "Foucault." And the director seems to delight in having the *favela* ganglord humiliate them, shoot them and light them on fire. (Indeed, in that film, Sandro's last name "Nasciemento" is given, discreetly, to the other side – the BOP commander who is at once ruthless and fascistic, hyped on drugs (like Sandro), and the nihilistic anchor of the sickness.) Padilha revolts – as does Sandro – by taking cinema hostage, putting it in hiatus, and identifying with what is nonetheless outside of the bus.

70 The term "biopolitics," commonly derived from Giorgio Agamben's appropriation of the term from Foucault, is that the modern management of *life* entails dividing the term into two zones – human life or that of the polis (*bios*) and "mere life" (*zoe*). The latter zone includes animals, things, organic process and disposable humans banned from the status of the living (like the street kids in *Bus 174*). Both Agamben and Žižek evoke the slums or *favelas* as an example of this "bare life" that is nonetheless taken into the polis. Padilha rewrites this divide, at first, as having to do with the visible and the "invisible." I will suggest that Padilha's *Bus 174* exceeds this first map altogether in confronting the global *telepolis*.

71 Žižek, "Nature and Its Discontents" 42.

72 These witnesses, speakers, cops and acquaintances arrive on screen as individuals yet assemble almost as TV types: the selfish aunt, the liberal sociologist, the bureaucratic SWAT leader, the masked gangster. The numerous speaking faces (some blurred or covered) form a sort of circuitry or transindividual display of what is called "Brazil."

73 By all measures, "Sandro" has become a sort of star. He has his own Wiki page and the film mentioned, *Ultima Paraida 174* (2008), is a feature film directed by Bruno Barreto in which signal features of Sandro's story are reshaped and dramatized. A morality tale, this full narrativization of Sandro can be seen as the opposite, rather than an extension of, what Padilha unleashes as his "Sandro."

74 The "interrupted robbery" (a robbery that, after all, is itself a pretext) is another covert name for cinema itself – and returns us to Benjamin's caesura, the cinematic "cut," and so on. It also implies the manner in which the image involves both the robbery effected by representation and the attempt to interrupt that itself as a mnemopolitics. In this hiatus, Sandro ceaselessly measures, defers and references time or the instant of killing (a mythic 6 pm). Padilha marks this in the entry to Nova Holanda slum in citing the "*Stop-Time Hotel*" on screen, where we are told street kids and gangsters used.

75 Padilha's *Tropa de elite* drew controversy for what seemed its identification with the brutal ("fascist") tactics of the BOPE. Yet in giving his BOPE chief the name "Nascimento," the same as Sandro, he indicates the *symmetrical* inversion of rhetorical positions concealed in the feature-film narrative.

76 *Sabotage* is cited in Tarantino's *Inglourious Basterds* (2009) with this same image in mind – that of the (cinematic) bus being exploded with a bomb associated with film itself. Tarantino appropriates the trope of a movie house as the spy front for blowing up itself with its audience. He applies this, wincingly, to "World War II" as what might be termed a cinematic war. *Sabotage*, however, continues to insinuate cinematics not only into history but into life and "human" form. It visits zoos, references animals for eating and associated with bombs (birds), passing to animation, a Disney cartoon of a half-human/half-bird in the film house. It opens a logic of cin-animation that exceeds Scotland Yard's policing of representation.

77 A recurrent covering, blotting, or *taking away of face* passes virally through different social types and figures – policemen, street kids, Sandro (and Padilha, who appears nowhere on the screen or in the presentation). This is a figure of invisible surveillance that not only incriminates the viewer but threads the numerous talking heads and the "omni" of the title – the "Brazil" of the screen. Such defacement moves toward a pre-individuated locus attached early on to the mountains, the water surface, and reappears at the end of *Tropa de elite* (2007), where the final wasting of the drug lord does buy blowing away his face (which he asks alone not be done) so SHOULD THIS BE DELETED? with a shot into the camera and a whiteout. *Defacement* is one of Padilha's elite tropes.

78 See Read 135.

79 See Taibbi, "Trump the Destroyer."

80 John Paczkowski, "Code/Red: The Best Elon Quote, Ever." https://www.vox.com/2014/10/1/11631474/codered-the-best-elon-musk-quote-ever : "Fuck Earth! Who cares about Earth?" Musk said. "If we can establish a Mars colony, we can almost certainly colonize

the whole solar system, because we'll have created a strong economic forcing function for the improvement of space travel...." Also: Ross Anderson, "Exodus," *Aeon*. https://aeon.co/essays/elon-musk-puts-his-case-for-a-multi-planet-civilisation.

Images

p. 26   Thick pools of oil on the water. Lousiana GOHSEP, Courtesy of Govenor Jindal's office, Copyright 2007, CC-BY-SA.

p. 29   A floating blob of oil from Deepwater Horizon. MC1 Jeffery Tilgman Williams, Public Domain.

p. 30   Chevron's Corporate Logo and tagline. Trademark Chevron Corporation. Photograph of a McDonald's hamburger. Evan Amos, Copyright 2014, CC-BY-SA. Wikimedia Commons.

p. 31   Oil for gold: Bond film, *Goldfinger*, updated later by *Quantum of Solace*. Still from the movie *Goldfinger*, Copyright 1964. Still from the movie *Quantum of Solace*, Copyright 2008.

SEM photography inducts new beings into its surveillance gallery. Photomicrograph of a Yellow Mite. Eric Ebe colorization Chris Pooley, Public Domain. Photomicrograph of a beetle. Gabriel Ewig, Copyright 2014, CC-BY.

p. 35   Still from the movie *Inglourious Basterds*, Copyright 2009.

p. 36   The opening of *Rear Window*, sleuthing lens inspects photos prospectively identifying the cause of James Stewart's immobilizing leg cast in a race-car crash, or the catastrophe of photography. Stills from the movie *Rear Window*, Copyright 1954.

p. 37   Linking the advent of photography (and cinema) to atom blast. Still from the movie *Rear Window*, Copyright 1954.

p 38   Shots of afflicted wildlife were iconic for the Gulf "leak." Lousiana GOHSEP, Courtesy of Govenor Jindal's office, Copyright 2007, CC-BY-SA.

p. 40   Altering the visual refraction of light. Deepwater Horizon oil slick. NASA Terra Satellite. Public Domain.

p. 42   Clumped by dispersants, oil in a rip tide. Louisiana Department of Wildlife and Fisheries, Copyright 2009, CC-BY-SA.

p. 43   Ingestion, incorporation. Deepwater Horizon. David L Valentine, Copyright, CC-BY-SA. Deepwater Horizon. Kris Krüg, Copyright 2010, CC-BY 2.0.

p. 44   Hands, technics, liquefaction. Oil covered hands. NARA, Public Domain. Hands dripping with oil, Unknown.

p. 45   US Military work to contain oil spill. Public Domain. New Harbour Island during the Deepwater Horizon spill, USGS, Public Domain.

p. 46   Not quite a face. Pelican covered in Oil. Lousiana GOHSEP, Courtesy of Govenor Jindal's office, Copyright 2007, CC-BY-SA. Turtle covered in oil. NOAA, Public Domain.

p. 47   Oil spill. NOAA, Public Domain. Deepwater Horizon spill. ESA TerraSAR-X, Copyright 2010, CC-BY.

p. 48   Transmutation to flames – "stored sunlight" returned. Deepwater Horizon oil rig on fire. US Coast Guard, Public Domain. Deepwater Horizon spill. NASA, Public Domain.

p. 50   Face of Norman – in "mother's voice" – dissolves into death grin imposed over bog, from which Marion's car is drawn by celluloid citing chain. Inverse phoenix. Still from the movie *Psycho*, Copyright 1960.

Inkpool of the (culpabale) judge's pen dissolves into a black drowning pool. Stills from the movie Manxman, Copyright 1929.

p. 51   Still from the movie *The Birds*, Copyright 1963.

p. 52   "Norman" before the bog, a set-up photographic shot, with looming ghost tree prop. Still from the movie set of *Psycho*, Copyright 1960.

p. 53   Still from the movie set of *Psycho*, Copyright 1960.

p. 56   Keiyo petrochemical complex. Nanashinodensyaku, Copyright 2014, CC-BY-SA.

p. 74   Smoke on the water. David L Valentine, Copyright, CC-BY-SA.

p. 90   Starling murmuration. Caroline Legg, Copyright, CC-BY.

p. 92   Myriad wings above Union Square, San Francisco, with its penchant for airline promotion. Still from the movie *The Birds*, Copyright 1963.

p. 100 Blind man of cinema – Dan Fawcett, eyes pecked out. Still from the movie *The Birds*. Copyright 1963.

p. 112 Palm oil plantations. ESA, Copyright, CC-BY-SA.

p. 136 Black and white rose, color inverted. Photospook, Copyright, CC-BY-SA.

p. 152 Law officers behind confiscated still. Unknown, Public Domain.

p. 153 Bearded man holding a gun and a pistol. Unknown.

p. 186 Reftinsky resevoir. Vasiliy Iakovlev, Copyright 2019, CC-BY.

p. 195 Lascaux horse replica. Patrock Janacek, Copyright 2012, CC-BY.

p. 196 Paper sky lanterns, inscribed with names and messages, cinematic figures yield to stunning galactic configuration from the expanse. Stills from the movie *Melancholia*, Copyright 2011.

p. 197 Two orbs momentarily echoing as eyes or reel – rogue unseen planet Melancholia and Earth – approach impact. Stills from the movie *Melancholia*, Copyright 2011.

p. 198 "What's the point?": wings, points, cuts. Still from the movie *The Birds*, Copyright 1963.

p. 199 Justine anti-grav. Still from the movie *Melancholia*, Copyright 2011.

p. 200 The "magic" cave, cinematic sanctuary. Still from the movie *Melancholia*, Copyright 2011.

p. 202 Final exodus of homeless Brenners – humans driven out. Still from the movie *The Birds*, Copyright 1963.

p. 214 Huricane Florence photographed from the International Space Station. NASA 2018, Public Domain.

p. 216 Super Mecha observing Boybot addressing Coney Island statue of "Blue Fairy". Still from the movie *A.I. Artificial Intelligence*, Copyright 2001.

p. 217 "Those were the years after the ice caps had melted because of the greenhouse gases, and the oceans had risen to drown so many cities along all the shoreliness of the world. Amsterdam, Venice, New York, forever lost…" Still from the movie *A.I. Artificial Intelligence*, Copyright 2001.

p. 218 Dr. Hobby's exemplary bot, "Sheila." Still from the movie *A.I. Artificial Intelligence*, Copyright 2001.

p. 220 Still from the movie *A.I. Artificial Intelligence*, Copyright 2001.

p. 221 David destroys his simulacrum double – while reading. Still from the movie *A.I. Artificial Intelligence*, Copyright 2001.

p. 222 De-extinct Monica in auteur David's mise-en-scène. Still from the movie *A.I. Artificial Intelligence*, Copyright 2001.

p. 223 Biomorphic Moebius graphics inhabiting and miming "eye." Still from the movie *Vertigo*, Copyright 1958.

p. 224 Monica with tell-tale spaghetti strand. Still from the movie *A.I. Artificial Intelligence*, Copyright 2001.

p. 225 David's face, inverting the "uncanny." Still from the movie *A.I. Artificial Intelligence*, Copyright 2001.

p. 227 David discovers the "David" factory. Still from the movie *A.I. Artificial Intelligence*, Copyright 2001.

p. 229 Dr. Know, CGI heir and update of *The 39 Steps'* Mr. Memory. Still from the movie *A.I. Artificial Intelligence*, Copyright 2001.

p. 230 Forest where David is abandoned like discarded pet by Monica. Still from the movie *A.I. Artificial Intelligence*, Copyright 2001.

p. 231 David's face melts when tricked into eating. Still from the movie *A.I. Artificial Intelligence*, Copyright 2001.

p. 232 Opening title sequence of *Vertigo* – moving from lips to darting paranoid eyes to entering the eye to encounter the spinning biomorphic bands that mutate space and time and post a graphic mobile within "eye," a woman's. Still from the movie *Vertigo*, Copyright 1958.

p. 233 Super Mecha observe revived Botboy David turned to the Coney Island statue of the "Blue Fairy" he invests magic hope in from reading Pinocchio. Still from the movie *A.I. Artificial Intelligence*, Copyright 2001.

p. 236 Rocinha Favela Brazil. Alicia Nijdam, Copyright, CC-BY.

p. 237 The invisible acrobats of Rio street kids. Still from the movie *Bus 174*, Copyright 2002.

p. 239 Acrobats of "invisibility" – Rio street kids. Still from the movie *Bus 174*, Copyright 2002.

p. 242 Sandro performs before the Globo camera. Still from the movie *Bus 174*, Copyright 2002.

p. 242 Hitchcock's *Sabotage* – perpetually mis-timed time-bomb blows up *bus* carrying it to Picadilly Circus ("the center of the world"). Still from the movie *Sabotage*, Copyright 1936.

p. 248 Inscriptions on wall of the vault, sunless storage cell. Still from the movie *Bus 174*, Copyright 2002.

p. 252 Final visit or descent to underworld of anonymous Brazilian prison presented like a negative drawing attention to the screen itself. Still from the movie *Bus 174*. Copyright 2002.

p. 253   "Stop Time Hotel" traversed by tele-wiring. Still from the movie *Bus 174*, Copyright 2002.

p. 254   Approach from the sea opening *Bus 174*, traversing media wired mountains encircling Rio, descending into favela and into downtown traffic, stuccoed with cameras. Stills from the movie *Bus 174*, Copyright 2002.

p. 260   Image generated by Stable Diffusion X with the prompt: The Trumpocene embraces climate breakdown, acclerating the bifucation between a hyper-enhanced techno-elite and an obsolete humanity. Public Domain.

www.ingramcontent.com/pod-product-compliance
Lightning Source LLC
Chambersburg PA
CBHW061249230426
43663CB00022B/2957